国家林业和草原局普通高等教育"十四五"规划教材

中华农耕文明

（上册）

胡家英　主编

中国林业出版社

图书在版编目（CIP）数据

中华农耕文明. 上册 / 胡家英主编. —北京：中国林业出版社，2022.7（2024.9 重印）
国家林业和草原局普通高等教育"十四五"规划教材
ISBN 978-7-5219-1696-6

Ⅰ. ①中… Ⅱ. ①胡… Ⅲ. ①农业—英语—高等学校—教材 Ⅳ. ①S

中国版本图书馆 CIP 数据核字（2022）第 085842 号

中国林业出版社·教育分社

策划、责任编辑：曹鑫茹　　　责任校对：苏　梅
电话：（010）83143560　　　传真：（010）83143516

出版发行	中国林业出版社（100009　北京市西城区刘海胡同 7 号）
	E-mail：jiaocaipublic@163.com　　电话：（010）83143500
	http：//www.forestry.gov.cn/lycb.html
印　刷	北京中科印刷有限公司
版　次	2022 年 7 月第 1 版
印　次	2024 年 9 月第 2 次印刷
开　本	787mm×1092mm　1/16
印　张	8.25
字　数	270 千字
定　价	36.00 元

未经许可，不得以任何方式复制或抄袭本书之部分或全部内容。

版权所有　侵权必究

《中华农耕文明》(上册)编写人员

主　编　胡家英

副主编　徐　虹

编　者　樊绪岩　于　洋　于　娜

序

Foreword

东北农业大学胡家英教授多年从事农业院校本科、硕士、博士英语教学，对涉农语言文化进行深入研究，为宣传中国传统农耕文化，编写了《中华农耕文明》（上、下册）。2021年8月教育部印发的《加强和改进涉农高校耕读教育工作方案》明确提出要加强学生传统农业文化教育，将耕读教育相关课程作为涉农专业学生必修课。《中华农耕文明》（上、下册）的出版，对弘扬我国耕读优秀传统文化，加强大学生传统农业文化教育，促进农业特色通识教育，培养"大国三农"情怀将起到积极作用。

《中华农耕文明》（上、下册）的编写坚持语言、文化、思维能力同步提高的教学理念，坚持课程思政，弘扬中国优秀传统文化，增强民族自信心，提升国际知名度，以中华农耕文明优秀传统文化为主线编写而成，是我国首部用英语系统讲授中华农耕文明的一套教材。该教材立意新颖、取材颇具匠心，从我国农业的起源，如农耕五祖、原始农业的刀耕火种，到农耕思想形成和古代农书编撰；从中国传统本土农作物的培育，如五谷和中药等，到传统农业技术文化，如土壤耕作制度、农具的革新和水利灌溉工程；从农时的推算，如24节气和72物候，到农耕风俗、农耕节日、农耕艺术和农耕谚语等方面介绍了中华农耕文明的发展。

农业院校的学生应该了解我国的农耕社会文明，这对深刻理解我国的政策方针和对今后的科研工作，有很积极的意义，尤其是学会用英语表达我国的优秀传统农耕文明，讲好中国农耕故事，至关重要。我祝愿胡家英教授团队在这个领域取得更大成绩，并希望该套教材在使用的过程中不断完善，成为一套优秀教材。

是为序。

罗锡文
中国工程院院士、华南农业大学教授
2022年6月18日

前 言
Preface

在绵延不断的历史长河中，炎黄子孙植五谷、饲六畜，农桑并举，耕织结合，形成了渔樵耕读、精耕细作、富国足民的优良传统，创造了上下五千年灿烂辉煌的中华文化。几千年来，中华农耕文明以其独特的文化价值和人文价值，向世界展示了中华文明的无穷魅力与风采。随着农业文明的不断发展，中华农耕文明成为中华民族一份沉甸甸的文化遗产。"耕读传家远，诗书继世长"，逐渐成为中国乡土文化的一大特色。

2021年8月，教育部印发《加强和改进涉农高校耕读教育工作方案》（以下简称《方案》）。《方案》明确提出加强学生传统农业文化教育，将耕读教育相关课程作为涉农专业学生必修课，编写中华农耕文明等教材，强化有关中华农耕文明、乡土民俗文化、乡村治理等课程教学。《中华农耕文明》（上、下册）旨在弘扬我国耕读传家优秀传统文化，加强大学生传统农业文化教育，以此促进农业特色通识教育课程体系建设，培养学生"大国三农"情怀。

本教材的编写坚持语言、文化、思维能力同步提高的教学理念，坚持课程思政，弘扬中国优秀传统文化，增强民族自信心，以内容教学法（Content-based Instruction，简称CBI）教学理念为理论依托，遵循以下三个基本原则。

一是以中华农耕文明优秀传统文化为学习主题进行语言训练的原则。本教材的编写旨在提升学生对中华农耕文明的认识和理解，并学会用英语表述，以此提升英语语言应用能力，达到用英语讲述中华农耕文明的目的，即中华农耕文明知识与语言工具兼得。

二是点面结合原则。《中华农耕文明》的编写主要从面上展示了中华农耕文明的历史面貌，从历史发展的纵向演变角度，展示农耕文明发生发展的历史脉络和标志性历史成就；《中华农耕文明》主要通过主题的拓展，从点上具体解读和展示中华农耕文明所包涵的内容及其相互关系。

三是创新原则。从内容方面，本教材比较全面地总结了中华农耕文明的相关事件和知识。从古代农学思想、精耕细作传统、农业技术文化、农业生产民俗、物候与节气文化、节庆文化、民间艺术、涉农诗词歌赋等方面概况介绍农耕文明的发展。从语言训练方面，本教材以英文为语言载体，聚焦中国优秀农耕文化知识解读，有助于中国文化的海外传播，课程思政贯穿始终。从练习的设计方面，体现读者的阅读体验和批判性思维，以独特的审美眼光指导教材的编写。从技术方面，本教材可用于线上线下混合课程，同步慕课已经全面上线。

《中华农耕文明》编写组成员主要来自东北农业大学教师团队。主编胡家英负责主题设计、编写规划等；于洋编写了上册第一章、第二章和第三章；徐虹编写了上册第四章、第五章和第六章；宋宝梅编写了下册第一章；樊绪岩编写了上册第七章、第八章，下册第二章、第三章、第四章和第五章；满盈编写了下册第六章、第七章和第八章；潘秋阳、于娜、岳欣、殷际文负责全书的校对工作。

本教材从设想到编写，从试用到出版，得到了许多同行、专家、教师和同学的支持和帮助。他们对全书的设计和编写给予了很多建议和支持。在此，我们对他们表示最衷心的感谢。

由于水平有限，缺失和不当在所难免，恳请各位同道与使用者批评与指正。

编　者

2021年12月

目 录
Contents

序
前 言

第一章　农业起源
Chapter 1　Origin of Agriculture ·· 1
　　Text A　农耕五祖　Five Farming Forefathers ······························· 1
　　Text B　长江流域的农业文化　Agriculture in the Yangtze River Valley ······ 8
　　Text C　原始农业的刀耕火种　Slash and Burn of Primitive Agriculture ····· 14

第二章　农耕思想
Chapter 2　Farming Ideology ·· 18
　　Text A　天人合一价值观　Values of Unity of Heaven and Humans ········ 18
　　Text B　和平主义价值观　Values of Pacifism ································ 26
　　Text C　家庭主义价值观　Values of Familism ······························· 30

第三章　传统农作物
Chapter 3　Traditional Cultivated Crops ·· 34
　　Text A　五谷　Five Cereals ·· 34
　　Text B　中药　Traditional Chinese Medicine ································ 42
　　Text C　园艺　Traditional Horticulture ······································ 46

第四章　土壤耕作制度
Chapter 4　Soil Cultivation System ·· 51
　　Text A　山脊种植和围垦　Ridge Cultivation and Polder ···················· 51
　　Text B　复种梯田　Multiple Cropping and Terraces ························ 56
　　Text C　精耕细作体系　Intensive Cultivation System ······················ 60

第五章　传统农业灌溉技术
Chapter 5　Irrigation ·· 65
   Text A　古代农业灌溉　Ancient Irrigation ·· 65
   Text B　旱地水资源保护　Water Conservation on Dry Land ··· 69
   Text C　灌溉工程　Famous Irrigation Projects ··· 74

第六章　传统农耕用具的技术革新
Chapter 6　Technological Innovations of Farming Implements ················· 79
   Text A　工具材料的创新　Innovation of Tool Materials ·· 79
   Text B　种植工具的创新　Innovation of Planting Tools ·· 83
   Text C　收获割工具的创新　Innovation of Harvesting Tools ··· 87

第七章　农耕时令——二十四节气
Chapter 7　Farming Time—The 24 Solar Terms ··· 92
   Text A　二十四节气　The 24 Solar Terms ·· 92
   Text B　二十四节气解说（1）　The Exact Interpretations of the 24 Solar Terms (1) ············· 98
   Text C　二十四节气解说（2）　The Exact Interpretations of the 24 Solar Terms (2) ············ 101

第八章　农耕时令——七十二物候
Chapter 8　Farming Time—The 72 Phenological Terms ···························· 106
   Text A　七十二物候　The 72 Phenological Terms ·· 106
   Text B　七十二物候解说　The Exact Interpretations of 72 Phenological Terms ·················· 111
   Text C　物候学　Phenology ·· 115

参考文献
Reference ··· 120

第一章 农业起源

Chapter 1 Origin of Agriculture

农耕五祖
Five Farming Forefathers

Chinese farming civilization developed long before written Chinese was invented. Various legends tell that the de facto agriculture in ancient China originated from about 5000 years ago. Legend has it that there arose five farming forefathers in Chinese erstwhile farming history, i.e. Shennong, Houji, Fuxi, Leizu, and Dayu. Their great feats go as follows. Shennong who had tasted a hundred of herbs, or even more, is honored as founding father of Chinese Herbal Medicine. It is Houji who, first and foremost, instructed people how to plant thus being revered as the farming forefather. Pioneering how to raise livestock, Fuxi is held in high esteem as the forefather of animal husbandry. Leizu taught people how to farm silkworm and thus is venerated as the forefather of sericulture. Saving Chinese people from flood and teaching them how to control the flood, Dayu is held in high reverence as the forefather of water conservancy. Thanks to their great wisdom, the tribal people in ancient China could never fall prey to wandering life (wearing animal's fur, eating animal's raw meat, fishing and hunting), but settle down under aegis of slash-and-burn, weaving and garment-making.

Shennong, Divine Farmer Tradition

Known as "Emperor of Five Grains" (五谷先帝, wugu xiandi), Shennong (神农, Divine Farmer or Spirit Farmer), was a legendary ruler of China and culture hero reputed to have lived during the time of Three Sovereigns and Five Emperors, some 5000 years ago. Shennong's most common name is made up of the characters for "god" or "deity," shen (神), and nong (农), which means "peasant" or "farmer". Therefore, Shennong literally means "farmer god". He's also known as wugushen (五谷神), the "God of Five Grains" or wuguxiandi, the "First God of the Five Grains". One of the most peculiar things about Shennong is that he's "bull-headed". In some artistic representations, he merely has horns or subtle bumps on his head, but in others, he literally has the head of a bull. Shennong is also said to have a forehead as hard as bronze, a skull as hard as iron, and a transparent stomach, which he used to observe how the herbs he ingested affected his body. He usually dresses in a simple robe made from leaves and foliage, sporting long hair and an overgrown beard, and is often depicted in his signature pose—sitting while munching on a branch. Shennong is among the group of variously named heroic persons and deities who have been traditionally given credit for various inventions: these include the hoe, plow (both leisi style and the plowshare), axe, digging wells, agricultural irrigation, preserving stored seeds by using boiled horse urine, the weekly farmers market, the Chinese calendar (especially the division into the 24 jieqi or solar terms), refinement of the therapeutic understanding of taking pulse measurements, acupuncture, and moxibustion, and institution of the harvest thanksgiving ceremony.

Shennong is one of the "the three great emperors" of Chinese legend. Shennong, as the Tianzu (田祖, field forefather) and Xiannong (先农, pioneering farmer), has been worshipped by the Chinese people in that he is the first progenitor who is engaged in farming practices. It is Shennong that taught people the skills of farming. He invented farming tools like lei and si and created means of farming, such as clearing wild land with fire, for which he got his nickname Lieshan (烈山氏), which literally means "burning the mountain to hunt". Thanks to his teachings, people gradually settled down and made their livings by farming.

Shennong is also said to be Father of Chinese herbal medicine. According to legend, his stomach was transparent and as a result he could distinguish different medicinal properties of herbs. While tasting and testing different herbs, he was frequently poisoned but luckily survived until a lethal plant finally claimed his life.

Shennong is also thought to be Yan Di (炎帝), the second of the three legendary kings of China. This title, however, is most commonly translated as "the Emperor of Fire". The Chinese character of yan (炎) is composed of two 火 which means fire. While Yan was not the ruler of the kingdom, people at that time worshiped him as an emperor because he taught his people how to use fire for farming.

Houji, Lord of Millet Tradition

Houji (后稷, Houji; literally Lord Millet) is a legendary Chinese cultural hero credited with introducing millet to humanity during the time of the Xia dynasty. Millet was the original staple grain of northern China, prior to the introduction of wheat. His name is translated as Lord of Millet and was a posthumous name bestowed on him by King Tang, the first of the Shang dynasty. Houji is worshipped for the abundant harvests that he graciously provided for his people. The Chinese honors him not only for past favors but in the hope that devotion to the deity would guarantee continued blessings.

Houji is another legendary figure who was born over 4000 years ago in the You Tai tribe. Houji, the successor of the Emperor Yan, is said to be born by a woman named Jiang Yuan who accidentally puts her foot on the footprint of the giant and then gives birth to him. One day, his mother, Jiang Yuan, saw a giant footprint while walking in the wild. Curiously, she stepped into it. Soon after that, she found herself pregnant. Taking it as an inauspicious or evil omen, the infant, Houji, was abandoned by Jiang on the narrow path of the hillside as soon as he was born. However, to Jiang's astonishment, every time Houji was deserted by her, Houji was brought back safe and sound. When Houji was placed on a road, animals surrounded him and protected him from being hurt; when on ice, birds warmed him with their feathers; when in a forest, woodcutters saved him. Houji's mother thought it might be god's will and finally she decided to keep her baby. Because of his experience of being deserted, thus the little infant Houji is also called Qi (弃) literally meaning being "abandoned" in Chinese. With farming being the daily work of the tribe at that time, Qi fell in love with agronomy when he was a little boy. Under the instruction of his mother, Qi quickly masters the agricultural knowledge. When Qi grew up, he repeatedly thought, observed, practiced, and taught people to grow the crops, spreading the farming culture. He cultivated grains like millet, wheat and soy beans that his tribe had never seen before. In addition, he taught the people other new practices including selecting seeds, plowing lands, removing weeds and using fertilizers, thus considerably boosting productivity. He also cultivated hemp of which the fiber could be made into clothes.

When Shun succeeded Yao as the Emperor of China, he appointed Qi as Houji, a position similar to Minister of Agricultural Affairs. Qi was so widely admired that Yao honored him with a surname Ji and the region where the You Tai tribe lived as his manor. For people at the time, it was a great honor to be given a surname by the emperor. Eventually, Qi became a great agronomist in ancient times and was revered as the farming forefather—Houji.

Fuxi, Pioneer of Livestock Tradition

Soon afterwards, there came a person named Fuxi who taught people how to knot the strings into a net to hunt birds and how to fish. Hence, the erstwhile primitive hunting of human beings entered into the primary animal husbandry production making outstanding contributions to the Chinese civilization. In Chinese mythology, Fuxi or Fu Hsi (伏羲, fuxi); aka Paoxi (庖牺), mid-2800s BC, was the first of the mythical Three Sovereigns (三皇, sanhuang) of ancient China. According to Chinese mythology, he is a cultural hero, reputed to have taught the Chinese people fishing with nets, hunting with weapons made of iron, cooking, domestication of animals, music, the writing system, sericulture (cultivation of silk worms) and the weaving of threads from silkworm cocoons into textiles. He tamed the waters of the Yellow River by digging dikes, canals and irrigation ditches, offered the first open air sacrifice and standardized marriage contracts. He is also credited with creating the eight trigrams (八卦), which form the basis for the philosophy of *The Book of Changes* (《周易》, *I Ching*, also known as the *Yijing* or *Zhouyi*) and are regarded as the origin of calligraphy. It is said that the Fuxi is featured with a human face but a snake body. His birth is extraordinarily legendary in that he was born after his mother had stepped on the footprints of a giant in a place called Leize (雷泽) and had been bearing him in pregnancy for 12 years.

Leizu, Inventor of Sericulture Tradition

Legend has it that the process for making silk cloth was first invented by the wife of the Yellow Emperor, Leizu, around the year 2696 BC. The idea for silk first came to Leizu while she was having tea in the imperial gardens. A cocoon fell into her tea and unraveled. She noticed that the cocoon was actually made from a long thread that was both strong and soft. Leizu then discovered how to combine the silk fibers into a thread. She also invented the silk loom that combined the threads into a soft cloth. Soon Leizu had a forest of mulberry trees for the silkworms to feed on and taught the rest of China how to make silk. Since silk fabric was invented in Ancient China by Leizu, it played an important role in their culture and economy for thousands of years.

Yu the Great, Forefather of Water Conservancy

Qi lived under the reign of Yao, an Emperor of ancient China, and during this time the empire was once struck by severe floods. The founder of the Xia Dynasty, Emperor Dayu was a historical figure who lived in the 23rd century BC. His father Gun had been commanded by Shun, the previous ruler, to control the Great Flood, but various Godly machinations prevented him doing so. Dayu, the dutiful son, felt honor-bound to take up the task in his place. Qi helped Yu, who was believed to be a descendant of Emperor Huang and was revered as Yu the Great, to control the floods and later was appointed by Yao to restore agriculture after the floods were finally tamed. For his unceasing labors, Dayu received popular acclaim and was rewarded with the title of Emperor. He is still revered to this day and honored as God of Flood Control, and his name is a byword for stupendous engineering projects.

Words & Expressions

erstwhile/'ɜːstwaɪl/*adj*. 以前的；从前的；往昔的
herb/hɜːb/*n*. 药草，香草；草本植物
esteem/ɪ'stiːm/*n*. 尊重，敬重；*v*. 尊重，敬重；认为，把…看作
sericulture/'serɪˌkʌltʃə/*n*. 养蚕；蚕丝业
feat/fiːt/*n*. 功绩，壮举；武艺，技艺；*adj*. 合适的，灵巧的
revere/rɪ'vɪə(r)/*vt*. 敬畏；尊敬；崇敬
esteem/ɪ'stiːm/*n*. 尊重，敬重；*v*. 尊重，敬重；认为，把…看作
venerate/'venəreɪt/*vt*. 崇敬，尊敬
tribal/'traɪb(ə)l/*adj*. 部落的，部族的；<贬>有组成集团倾向的，集团意识强的；部落的，部族的；*n*. （尤指印度次大陆的）部落成员（tribals）
aegis/'iːdʒɪs/*n*. 羊皮盾（宙斯及其女儿雅典娜所持的帝盾）
progenitor/prəʊ'dʒenɪtə(r)/*n*. 祖先；原著；起源

literally/'lɪtərəli/*adv*. 按照字面意义地,逐字地;真正地,确实地;(用于夸张地强调)简直

survive/sə'vaɪv/*v*. (经历事故、战争或疾病后)活下来,幸存;比…活得久,比…长寿;挺过,艰难度过;幸免于难,留存;(靠很少的钱)继续维持生活;设法对付(困难或令人不愉快的事)

lethal/'li:θ(ə)l/*adj*. 致命的,致死的;*n*. 致死因子

inauspicious/ˌɪnɔː'spɪʃəs/*adv*. 不吉利;不祥

omen/'əʊmən/*n*. 预兆,征兆;预兆性;*vt*. 预示,预告;占卜

desert/'dezət/*n*. 沙漠,荒漠;荒凉的地方;应得的赏罚;*v*. 离弃,舍弃(某地);抛弃,遗弃(某人);背弃,放弃;擅离,开小差;突然丧失;*adj*. 无人居住的,荒凉的;像沙漠的

agronomy/ə'grɒnəmi/*n*. 农学;农艺学;农业经济学

agricultural/ˌægrɪ'kʌltʃərəl/*adj*. 农业的,与农业有关的;务农的,农用的

fertilizer/'fɜːtəlaɪzə(r)/*n*. 肥料,化肥;受精媒介物

considerably/kən'sɪdərəbli/*adv*. 相当地;非常地

boost/buːst/*v*. 使增长,推动;[美,非正式]偷窃;宣扬,推广;*n*. 推动,促进

cultivate/'kʌltɪveɪt/*v*. 开垦,耕作;栽培,培育;陶冶,培养;建立(友谊)

hemp/hemp/*n*. 大麻;麻类植物;大麻烟卷;*adj*. 大麻类植物的

manor/'mænə(r)/*n*. 庄园;领地;采邑,采地

unravel/ʌn'rævl/*vt*. 解开;阐明;解决;拆散;*vi*. 解决;散开

mulberry/'mʌlbəri/*n*. 桑树;桑葚,桑椹;深紫红色

extraordinarily/ɪk'strɔːdnrəli/*adv*. 极其,极端地;奇怪地

de facto *adj*. [法]实际上的

first and foremost 首先;首要的是

under the aegis of…在…庇护/支持下,由…主办

thanks to 由于,幸亏

slash-and-burn *adj*. 刀耕火种的

give birth to 产生,造成;生孩子

safe and sound 安然无恙

tame/teɪm/*adj*. 驯服的,不怕人的;平淡无奇的,枯燥乏味的;听使唤的,温顺的;[美](植物)经栽培的;(土地)经开垦的;*v*. 驯化,驯服;制服,控制;开垦,开辟;

stupendous/stjuː'pendəs/*adj*. 惊人的;巨大的

Notes

1. lei si:耒耜,象形字,中国古代的像犁的一种翻土农具,形如木叉,上有曲柄,下面是犁头,用以松土,可看作犁的前身。"耒"是汉字部首之一,与原始农具或耕作有关。耒耜的发明开创了中国农耕文化。耜用

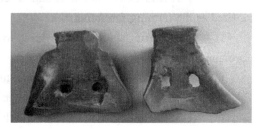

来起土，耒是耜上的木曲柄。耒耜，也用作农具的统称。

2. You Tai tribe：犹太人（希伯来语：יהודים，Jews，Yhudim），又称犹太民族，是广泛分布于世界各国的族群。根据犹太教律法《哈拉卡》的定义，一切皈依犹太教的人以及由犹太母亲所生的人都属于犹太人。犹太人发源于西亚的以色列地或希伯来地，犹太人的民族、文化和宗教信仰之间具有很强的关联性，犹太教是维系全体犹太人之间认同感的传统宗教。以色列是世界上唯一一个以犹太人为主体民族的国家。

3. Jiang Yuan：姜嫄，姓姜名嫄，有邰氏部落（今陕西眉县邰亭）人。上古时期历史人物，帝喾元妃，周族始祖后稷之母。

4. The eight trigrams：八卦，见于《周易·系辞下》云："古者包牺氏之王天下也，仰则观象于天，俯则观法于地；观鸟兽之文与地之宜；近取诸身，远取诸物，于是始作八卦，以通神明之德，以类万物之情。"八卦生自太极、两仪、四象中，"四象生八卦"。

5. *The Book of Changes*（《周易》）：《周易》包含《易经》，《三易》之一（另有观点：认为易经即三易，而非周易），是传统经典之一，相传系周文王姬昌所作，内容包括《经》和《传》两个部分。《经》主要是六十四卦和三百八十四爻，卦和爻各有说明（卦辞、爻辞），作为占卜之用。《传》包含解释卦辞和爻辞的七种文辞共十篇，统称《十翼》，相传为孔子所撰。

Reading Comprehension

Ⅰ. **Each piece of the following information is given in one of the paragraphs in the passage. Identify the paragraph from which the information is derived and put the corresponding number in the space provided.**

(　　) 1. A cocoon fell into her tea and unraveled.

(　　) 2. His birth is extraordinarily legendary in that He was born after his mother had stepped on the footprints of a giant in a place called Leize and had been bearing him in pregnancy for 12 years.

(　　) 3. He also cultivated hemp of which the fiber could be made into clothes.

(　　) 4. According to legend, his stomach was transparent and as a result he could distinguish different medicinal properties of herbs.

(　　) 5. Qi lived under the reign of Yao, an Emperor of ancient China, and during this time the empire was once struck by severe floods.

Ⅱ. **Decide whether the statements are true (T) or false (F) according to the passage.**

(　　) 1. It is Fuxi who, first and foremost, instructed people how to plant thus being revered as the farming forefather.

(　　) 2. Leizu taught people how to farm silkworm and thus is venerated as the forefather of sericulture.

(　　) 3. It is said that the Houji is featured with a human face but a snake body.

(　　) 4. Houji is also said to be Father of Chinese herbal medicine.

(　　) 5. Shennong is one of the "the three great emperors" of Chinese legend.

Language Focus

III. Complete the sentences with the correct form of the words in the table.

feat	Pioneer	extraordinary	agriculture	astonish
agronomy	revere	arise	venerate	auspicious

1. Legend has it that there _____ five farming forefathers in Chinese erstwhile farming history, i. e. Shennong, Houji, Fuxi, Leizu, and Dayu, from China or from India?

2. Leizu taught people how to farm silkworm and thus is _____ as the forefather of sericulture.

3. His birth is _____legendary in that He was born after his mother had stepped on the footprints of a giant in a place called Leize and had been bearing him in pregnancy for 12 years.

4. Taking it as an _____ or evil omen, the infant, Houji, was abandoned by Jiang on the narrow path of the hillside as soon as he was born.

5. However, to Jiang's _____, every time Houji was deserted by her, Houji was brought back safe and sound.

6. Eventually, Qi became a great _____ in ancient times and was revered as the farming forefather—Houji.

7. Under the instruction of his mother, Qi quickly masters the_____ knowledge.

8. _____ how to raise livestock, Fuxi is held in high esteem as the forefather of animal husbandry.

9. It is Houji who, first and foremost, instructed people how to plant thus being _____ as the farming forefather.

10. Their great _____ go as follows.

IV. Match the sentences in Section A with the English translation in Section B.

Section A

1．传说伏羲氏人面蛇身。
2．中国汉字——炎，是由两个"火"字组成，表示"火焰"。
3．嫘（léi）祖发明养蚕，被尊称为"蚕祖"。
4．弃成为远古时一位大农艺师，被尊称为农业始祖后稷。
5．神农，是我国传说中"三皇"之一。

Section B

1. Eventually, Qi became a great agronomist in ancient times and was revered as the farming forefather—Houji.

2. Shennong is one of the "the three great emperors" of Chinese legend.

3. It is said that the Fuxi is featured with a human face but a snake body.

4. Leizu taught people how to farm silkworm and thus is venerated as the forefather of sericulture.

5. The Chinese character of Yan (炎) is composed of two 火 which means fire.

V. Translate the paragraph into Chinese.

Chinese farming civilization developed long before written Chinese was invented. Various legends tell that the de facto agriculture in ancient China originated from about 5000 years ago. Legend has it that there arose five farming forefathers in Chinese erstwhile farming history, i.e. Shennong, Houji, Fuxi, Leizu, and Dayu. Their great feats go as follows. Shennong who had tasted a hundred of herbs, or even more, is honored as founding father of Chinese Herbal Medicine. It is Houji who, first and foremost, instructed people how to plant thus being revered as the farming forefather. Pioneering how to raise livestock, Fuxi is held in high esteem as the forefather of animal husbandry. Leizu taught people how to farm silkworm and thus is venerated as the forefather of sericulture. Saving Chinese people from flood and teaching them how to control the flood, Dayu is held in high reverence as the forefather of water conservancy. Thanks to their great wisdom, the tribal people in ancient China could never fall prey to wandering life (wearing animal's fur, drinking animal's blood, fishing and hunting), but settle down under aegis of slash-and-burn, weaving and garment-making.

_____

_____

_____

_____

Development

VI. Discuss the following questions.

1. Please find more English information about Fuxi to read and translate into Chinese, then think about his role in Chinese farming civilization.

2. Please find more English information about Leizu to read and translate into Chinese, then think about her role in Chinese farming civilization.

3. Please find more English information about Dayu to read and translate into Chinese, then think about his role in Chinese farming civilization.

长江流域的农业文化
Agriculture in the Yangtze River Valley

While cultures evolved and became prosperous one by one along the Yellow River, the Yangtze River also gave birth to cultures of her own. In 1936 a type of culture different from the

Yangshao Culture and the Longshan Culture was discovered in Liangzhu County of Hangzhou City, and as its first excavation was made in that place it is called the Liangzhu Culture. It started, according to C-14, about 3300 BC and ended about 2000 BC, and its domain had reached northwards to the Yellow River Valley and westwards to Anhui Province and Jiangxi Province with its core region being in Taihu Basin. During its span of over one thousand years it achieved no less than its contemporaries in the Yellow River Valley in building, agriculture, pottery making, jade ware making and other aspects. Up to now a city of more than three square kilometers has been excavated in its domain, which is the biggest of all that have been excavated belonging to this time in both the Yangtze River Valley and the Yellow River Valley. Remains of earth pyramids are also unearthed with the largest one having a base of about 10000 square meters. As for agriculture, people of this culture cultivated rice as well as vegetables and fruits. A piece of silk fabric is found among the relics dating back to about 3200—2700 BC. This is the most ancient silk fabric that has ever been found in China and in the world. Symbol of Liangzhu Culture may be yucong, a kind of jade ware being used for religious activity, the biggest of which is 31 cm high with the length and width being 7.5 cm. Some experts believe that Liangzhu Clan is no other than the powerful kingdom which is mentioned in an ancient book called *He Guan Tzu*, and that its people are the ancestors of later State Wu and State Yue in the Zhou Dynasty.

The Hemudu Culture is another one discovered within the Yangtze River Valley, whose duration is from about 5000 BC to about 3300 BC. Its domain barely confines to the Ningshao Plain which is between Ningbo City, Shaoxing City and northwards to the Hangzhou Bay. The excavations show that people of this culture had generally used leisi, a kind of primitive plough made of stone, therefore their agriculture might be prosperous. This is proved by a great amount of rice unearthed at one of its sites. To adapt to the southern climates and wet natural environments Hemudu people built their houses on wooden platforms above the ground.

Putting together the two cultures in the Yangtze River Valley, we find that at the time knot of 3300 BC the Hemudu Culture ended and the Liangzhu Culture began. Considering that they are geographically close to each other, a question may be put forward: did Liangzhu Culture grow from Hemudu Culture? To answer this question we have to wait for more archaeological discoveries.

Apart from all we have mentioned above, another culture being worth mentioning is the Hongshan Culture in the Liaohe River Valley of North east of China, which is confirmed to exist from about 5500 BC to about 3000 BC. Its outstanding excavations are Yuzhulong, a jade carving dragon with a round shape, which is hailed as the First China Dragon for its ancientness and elegance, and a beautiful head of a goddess stature, which indicates its high level of sculpting skill and prosperous religious activities.

To sum up all the unearthed relics of the Neolithic Age within China, we find that, compared with that of the Paleolithic Age, cultures of this period developed greatly. In respect of residence, people of this age lived no longer in caves or simple shelters but in well-made houses, even palaces. As for social form, people lived not only with their families but in big communities like village, city and tribe. Regarding their thought and conception, from their tombs, pictures on wares and some special buildings we get the information that they had primitive believes and even philosophical ideas. And primitive characters were also found on pottery pieces, some of them being arranged in lines. What's more, mental tools began to be made and used in this age. For instance, at Maojiayao Village, Dongxiang Country, Gansu Province, a bronze knife was excavated, which dates back to about 3000 BC. Up to the end of last century there had been at least mare than 30 bronze or pure copper tools to be unearthed including ax, chisel, knife, awl, dagger and mirror.

We cannot introduce all the prehistoric cultures discovered in China in this section. By the ones above we were able to have a brief tour to the Chinese prehistoric culture. Recapitulating what we have discussed, we get the main points as following:

Firstly, around the third millennium before Christ cultural prosperity appeared along the Yellow River, the Yangtze River and the Liaohe River in China.

Secondly, these Cultures belong to agricultural society with resident life. This is the key step for Chinese ancestors to transform from barbarism to civilization, as only agriculture and ecesis life can lay a solid foundation for an advanced civilization in a later time. The cultivation of rice and maize is the greatest contribution made by Chinese ancestors to the world agricultural civilization.

Thirdly, people of these cultures generally made and used pottery wares. Some of them are so fine that they are considered to prefigure the later invention of porcelain. So the New Stone Age or the Neolithic Period in China is in fact the "pottery age".

Fourthly, a great number of jade wares are excavated, which is seldom found in other prehistoric cultural regions outside China like Mesopotamia, the Nile Valley and the Indus Valley, showing that China has a long history of jade-use and jade-worship culture.

Fifthly, the silk fabric found in Liangzhu Culture shows that China's sericulture and textile are far earlier than any other country in the world and extends China's silk history to 5000 years.

Finally, excavations of bronze tools and wares at different sites of different cultures belonging to this age refute what Stavrianos, an American historian, indicates in his A Global History that China's use of bronze ware was started by a small group of invaders from the northwest, who brought the bronze-smelting technology into China around 1500 BC.

Words & Expressions

prosperous/'prɒspərəs/*adj*. 繁荣的，富足的
excavation/ˌekskə'veɪʃ(ə)n/*n*. 挖掘，发掘
domain/də'meɪn; dəʊ'meɪn/*n*. 领域，范围；领土，势力范围；（因特网上的）域；（函数的）定义域
therefore/'ðeəfɔː(r)/*adv*. 因此，所以
adapt/ə'dæpt/*v*. 适应；调整，使适合；改编；改造，改装
geographically/ˌdʒiːə'græfɪkli/*adv*. 在地理上，地理学上
archaeological/ˌɑːkɪə'lɒdʒɪk(ə)l/*adj*. [古] 考古学的；[古] 考古学上的

outstanding/aʊt'stændɪŋ/*adj*. 杰出的，优秀的；显著的，突出的；未解决的，未完成的；（款项）未支付的，未结清的

hail/heɪl/*v*. 赞扬，欢呼；呼喊，招呼；下冰雹；（大量物体）像雹子般落下（或击打）；[古] 表示问候或欢呼欢迎；*n*. 冰雹；雹子般的一阵；招呼（某人）

stature/'stætʃə(r)/*n*. 身高，身材；（精神、道德等的）高度

sculpting/'skʌlptɪŋ/*n*. 雕塑法；*v*. 雕刻；雕塑

relics/'relɪks/*n*. 遗迹；遗骸；纪念物（relic 的复数）

paleolithic/ˌpæliə'lɪθɪk/*adj*. 旧石器时代的

residence/'rezɪdəns/*n*. 住宅，住所；居住，定居；（在某国的）居住权，居留许可；（大学的）学生宿舍楼

bronze/brɒnz/*n*. 青铜；青铜色，古铜色；青铜艺术品；铜牌；*adj*. 青铜制的；青铜色的，古铜色的；*v*. 使被太阳晒黑；镶上青铜面

chisel/'tʃɪz(ə)l/*n*. 凿子；*v*. 雕，刻，凿；（非正式）欺骗

awl/ɔːl/*n*. 锥子；尖钻

dagger/'dægə(r)/*n*. 匕首，短剑；*vt*. 用剑刺；*n*.（Dagger）人名；（俄）达格尔

recapitulate/ˌriːkə'pɪtʃuleɪt/*vi*. 概括；重述要点；摘要；*vt*. 重述要点；摘要

millennium/mɪ'leniəm/*n*. 千年期（尤指公元纪年）；千周年纪念日；新千年开始的时刻

barbarism/'bɑːbərɪzəm/*n*. 野蛮；原始；未开化；暴虐

ecesis/ɪ'siːsɪs/*n*. [生态] 定居

maize/meɪz/*n*. 玉米；黄色，玉米色

prefigure/ˌpriː'fɪɡə(r)/*vt*. 预示；预想

porcelain/'pɔːsəlɪn/*n*. 瓷；瓷器；*adj*. 瓷制的

Neolithic/ˌniːə'lɪθɪk/*adj*. [古] 新石器时代的；早先的

no less than 不少于；多达…；正如

date back to 追溯到；从…开始有

apart from 远离，除…之外；且不说；缺少

in respect of 关于，涉及

machination/ˌmæʃɪ'neɪʃn/*n*. 阴谋；诡计

Notes

1. Yangshao Culture：仰韶文化，是指黄河中游地区一种重要的新石器时代彩陶文化，其持续时间大约在公元前 5000 年至前 3000 年（即距今约 7000 年至 5000 年，持续时长 2000 年左右），分布在整个黄河中游从甘肃省到河南省之间。因 1921 年首次在河南省三门峡市渑池县仰韶村发现，故按照考古惯例，将此文化称之为仰韶文化。其以渭、汾、洛诸黄河支流汇集的关中豫西晋南为中心，北到长城沿线及河套地区，南达湖北西北，东至河南东一带，西到甘肃、青海接壤地带。

2. Longshan Culture：良渚文化分布的中心地区在钱塘江流域和太湖流域，而遗址分布最密集的地区则在钱塘江流域的东北部、东部。该文化遗址最大特色是所出土的玉器。挖掘自墓葬中的玉器包含有璧、琮、冠形器、玉镯、柱形玉器和玉钺等诸多器型。

此外，良渚陶器也相当细致。不少学者认为良渚文化可以算得上中国的第一个王朝，据《韩非子·显学》，虞代延续了 1000 余年，推测良渚文化便是虞朝的考古学文化。陈剩勇、吕琪昌等强调良渚文化是夏文化的源头。

3. Yucong：玉琮是一种内圆外方筒型玉器，是古代人们用于祭祀神祇的一种礼器。距今约 5100 年至新石器中晚期，玉琮在江浙一带的良渚文化、广东石峡文化、山西陶寺文化中大量出现，尤以良渚文化的玉琮最发达，出土与传世的数量很多。

4. Hemudu Culture：河姆渡文化，是指中国长江流域下游以南地区古老而多姿的新石器时代文化（即距今约 7000 年前）。黑陶是河姆渡陶器的一大特色；在建筑方面，遗址中发现大量"干阑式房屋"的遗迹。

5. Hongshan Culture：红山文化，发源于东北地区西南部。起始于距今 5000～6000 年，是华夏文明最早的遗迹之一，分布范围在东北西部的热河地区，北起内蒙古中南部地区，南至河北北部，东达辽宁西部，辽河流域的西拉木伦河和老哈河、大凌河上游。

6. Yuzhulong：玉猪龙是我国对发现于红山等地的一种玉器的称呼，又名玉兽玦。被认为是龙的最早雏形。

7. The Paleolithic Age：新石器时代，是考古学家设定的一个时间区段，大约从 10000 年前开始，结束时间从距今 5000 多年至 2000 多年。

8. Mesopotamia：美索不达米亚平原，在两河流域。是位于底格里斯河、幼发拉底河之间的平原，现今的伊拉克境内，那里是古代四大文明的发源地之一古巴比伦所在，有高度发达的文明。 两河流域是世界上文化发展最早的地区，为世界发明了第一种文字——楔形文字，建造了第一个城市，编制了第一种法律，发明了第一个制陶器的陶轮，制定了第一个 7 天的周期，第一个阐述了上帝以 7 天创造世界和大洪水的神话。为世界留下了大量的远古文字记载材料（泥板）。

9. The Nile Valley：尼罗河流域分为七个大区：东非湖区高原、山岳河流区、白尼罗河区、青尼罗河区、阿特巴拉河区、喀土穆以北尼罗河区和尼罗河三角洲。英国探险家约翰·亨宁·斯皮克 1862 年 7 月 28 日发现了尼罗肥沃河在维多利亚湖的"源头"，该河北流，经过坦桑尼亚、卢旺达和乌干达，从西边注入非洲第一大湖维多利亚湖。尼罗河干流就源起该湖，称维多利亚尼罗河。河流穿过基奥加湖和艾伯特湖，流出后称艾伯特尼罗河，该河与索巴特河汇合后，称白尼罗河。另一条源出中央埃塞俄比亚高地的青尼罗河与白尼罗河在苏丹的喀士穆汇合。

10. The Indus Valley：印度河流域介于北纬 24°～37°，东经 66°～82°。东北倚喀喇昆仑山脉和喜马拉雅山脉，东南介印度塔尔沙漠，西北为阿富汗兴都库什山脉，西南为俾路支高原，南临阿拉伯湾。印度河流域属于亚热带气候，具有明显的季风气候特点，但由于东北部高山山脉的影响，使气候通常介于干燥与半干燥、热带与亚热带之间。一年分为四季。印度河灌溉历史悠久。早在公元前 3000 年，沿印度河两岸狭小的地带就已发展引洪灌溉。到公元五六世纪发展到修建引水灌溉渠道。

11. Stavrianos：斯塔夫里阿诺斯，美国历史学家，希腊族，学者、教授，出生于加拿大温哥华，毕业于不列颠哥伦比亚大学，在克拉克大学获文科硕士学位和哲学博士学位；曾任美国加利福尼亚大学的历史教授、西北大学的荣誉教授和行为科学高级研究中心的研究员。

12. *A Global History*：《全球通史》，由剑桥大学、牛津大学以及世界各国名牌大学史学专家合力创作，全景描述全球历史上发生的重大事件，全面展示世界悠久的历史和灿烂的文明。

Reading Comprehension

I. Each piece of the following information is given in one of the paragraphs in the passage. Identify the paragraph from which the information is derived and put the corresponding number in the space provided.

(   ) 1. Secondly, these Cultures belong to agricultural society with resident life.

(   ) 2. at the time knot of 3300 BC the Hemudu Culture ended and the Liangzhu Culture began.

(   ) 3. As for agriculture, people of this culture cultivated rice as well as vegetables and fruits.

(   ) 4. This is proved by a great amount of rice unearthed at one of its sites.

(   ) 5. Fifthly, the silk fabric found in Liangzhu Culture shows that China's sericulture and textile are far earlier than any other country in the world and extends China's silk history to 5000 years.

II. Decide whether the statements are true (T) or false (F) according to the passage.

(   ) 1. Liangzhu Culture started, according C-14, about 3300 BC and ended about 2000 BC.

(   ) 2. As for agriculture, people of Liangzhu Culture cultivated rice as well as vegetables and wheat.

(   ) 3. The excavations show that people of this culture had generally used leisi, a kind of primitive plough made of stone, therefore their agriculture might not be prosperous.

(   ) 4. The cultivation of rice and maize is the greatest contribution made by Chinese ancestors to the world agricultural civilization.

(   ) 5. the silk fabric found in Liangzhu Culture shows that China's sericulture and textile are far earlier than any other country in the world and extends China's silk history to 6000 years.

Development

III. Discuss the following questions.

1. Can you find more agricultural information about cultures in the Yangtze River Valley?

2. Can you find more agricultural information about cultures in the HuangHe River Valley?

Text C 原始农业的刀耕火种
Slash and Burn of Primitive Agriculture

The so-called slash-and-burn method refers to the action that mountain dwellers cut down trees at the first sign of spring and let them dry in the sun. Then at night before the spring rains, farmers set fire to the forest, burning down everything for fertilizer. Seeds were sown the next day while the soil was sill warm. Finally, without any additional soil management, farmers waited for the harvest. However, one drawback to the system was that usually after two or three years, the soil was no longer fertile enough to grow plants.

The first step during the slash and burn phase was to select the farming land. Based on the experiences of southern minorities during the early primitive agricultural period, the preferred land was woodland. The elderly of Dulong, Nu, Wa ethnic groups recalled that their ancestors selected forest outskirts, open space or sparse woodlands as farming land on the basis of tree type instead of soil property. They chose sunlit spots or flat areas (lower in the middle).

Step two: Decide the way to plant crops according to tree types. In the Dulong ethnic areas, people classified woodland into lumber land, bamboo land and mixed land. In lumber land where silei, simo grew, it was better to grow buckwheat, millet and barnyard grass; where ermang and jiu grew, it was better to grow corn; where wild walnut tree grew, it was better to grow taro. Mixed land was ideal for corn and millet. According to the Nu ethnic people's experience, the most suitable arable land was the place chestnut tree grew. Thus, during the time of slash and burn, to distinguish between varieties of botanic species of woodland was the foundation for land selection. This has been retained in the traditions of agriculture today. *Yin and Yang Miscellanea* matched five grains with five woods one by one. Rice depended on jujube or poplar; broomcorn millet depended on elm; soy bean depended on Chinese scholar-tree; red bean depended on plum; hemp depended on poplar or vitex; barley depended on apricot; wheat depended on peach; rice depended on willow or poplar.

The most typical production form of slash-and-burn was non cropping rotation, e.g. a patch of land was cultivated for one season only then left alone for natural rejuvenation for as long as 10 years. Lots of villagers nicknamed this sort of land as "lazy land". When processing the burning part of the method, it was not to aimlessly setting fire to primitive forest, but finished by meticulous plan. For example, in a village, villagers divided all the "lazy land" into ten sections and planted one section per year. Before setting fire to the lands, it was necessary to prepare the fire lane and install special guards to prevent cross-border fires. When trees were felled, the stumps of big trees as well as the roots of small trees were to remain in preparation for the next spring.

Words & Expressions

dweller/'dwelə(r)/*n.* 居民，居住者

drawback/'drɔːbæk/*n.* 缺点，不利条件；出口退税

fertile/'fɜːtaɪl/*adj.* 肥沃的，富饶的；可繁殖的，能结果的；能产生好结果的，促进的；点子多的，想像力丰富的；(核材料) 能产生裂变物质的，增殖性的

minorities/maɪ'nɒrətɪz/*n.* 少数 (minority 的复数形式)；少数民族；少数族裔

ethnic/'eθnɪk/*adj.* (有关) 种族，民族的；少数民族的；具有民族特色的，异国风味的；既非基督教亦非犹太教的，异教徒的；*n.* 少数民族成员，某民族群体的人

outskirt/'aʊtskɜːt/*n.* 郊区，市郊

lumber/'lʌmbə(r)/*v.* 笨重地行动，缓慢地移动；(非正式) 拖累；伐木；乱堆；逢场作戏地结识 (可能成为性伴侣)；迫使担负 (职责等)；烦扰；*n.* 木材；废旧家具，废物，无用的杂物

buckwheat/'bʌkwiːt/*n.* 荞麦；荞麦粉；荞麦片；*adj.* 蓼科植物的

taro/'tɑːrəʊ; 'tærəʊ/*n.* 芋头，[园艺] 芋艿

arable/'ærəb(ə)l/*adj.* 适于耕种的；可开垦的；*n.* 耕地

chestnut/'tʃesnʌt/*n.* 栗子；栗色；[园艺] 栗树；栗色马；*adj.* 栗色的

distinguish/dɪ'stɪŋgwɪʃ/*v.* 使有别于；看清，认出；区别，分清

botanic/bə'tænɪk(ə)l/*adj.* 植物 (学) 的；*n.* 植物制剂

miscellanea/ˌmɪsə'leɪniə/*n.* 杂集；杂录 (miscellany 的复数)

jujube/'dʒuːdʒuːb/*n.* 枣子，枣树；枣味糖

poplar/'pɒplə(r)/*n.* 白杨；白杨木

broomcorn/'bruːmˌkɔːn/*n.* [作物] 高粱

elm/elm/*n.* [林] 榆树；榆木；*adj.* 榆树的；榆木的

scholar-tree 槐树

vitex 牡荆属

barley/'bɑːli/*n.* 大麦

apricot/'eɪprɪkɒt/*n.* 杏，杏子；[园艺] 杏树；杏黄色；*adj.* 杏黄色的

cropping rotation 轮作

rejuvenation/rɪˌdʒuːvə'neɪʃn/*n.* [地质][水文] 回春，返老还童；复壮，恢复活力

install/ɪn'stɔːl/*v.* 安装，设置；(计算机) 安装 (新软件)；正式任命，使正式就职；安顿，安置

no longer 不再

set fire to 点燃，使燃烧

ethnic group 同种同文化之民族

Notes

1. **Dulong ethnic groups**：独龙族是中国人口较少的少数民族之一，也是云南省人口

最少的民族，使用独龙语，没有本民族文字。

2．Nu ethnic groups：怒族是中国人口较少、使用语种较多的民族之一，云南省怒江傈僳族自治州的泸水（原碧江县）、福贡、贡山独龙族怒族自治县、兰坪白族普米族自治县，以及迪庆藏族自治州的维西县和西藏自治区的察隅县等地。

3．Wa ethnic groups：佤族，中国、缅甸的少数民族之一，民族语言为佤语，属南亚语系孟高棉语族佤德语支，没有通用文字，人们用实物、木刻记事、计数或传递消息。

4．Silei：斯雷，树名，罕见，生长于独龙族地区。

5．Simo：斯莫，树名，罕见，生长于独龙族地区。

6．Ermang：尔芒，树名，罕见，生长于独龙族地区。

7．Jiu：纠，树名，罕见，生长于独龙族地区。

8．*Yin and Yang Miscellanea*：《杂阴阳书》或《阴阳杂技》，是汉代阴阳家著作。

9．Non-cropping rotation：无轮作轮歇类型，是最典型的刀耕火种形态，一块地只种一季就抛荒休闲，休闲期长达十年左右。

这种类型的地被很多民族称为"懒活地"（lazy land），意思是不需要怎么费劲儿，就可以获得收成，所以是各个民族的首选。

Reading Comprehension

I. Each piece of the following information is given in one of the paragraphs in the passage. Identify the paragraph from which the information is derived and put the corresponding number in the space provided.

(   ) 1. The first step during the slash and burn phase was to select the farming land.

(   ) 2. Mixed land was ideal for corn and millet.

(   ) 3. Before setting fire to the lands, it was necessary to prepare the fire lane and install special guards to prevent cross-border fires.

(   ) 4. decide the way to plant crops according to tree types.

(   ) 5. Seeds were sown the next day while the soil was sill warm.

II. Decide whether the statements are true (T) or false (F) according to the passage.

(   ) 1. However, one drawback to the system was that usually after two or three years, the soil was no longer fertile enough to grow plants.

(   ) 2. Based on the experiences of southern minorities during the early primitive agricultural period, the preferred land was grassland.

(   ) 3. In the Dulong ethnic areas, people classified woodland into lumber land, bamboo land and Thailand.

(   ) 4. Mixed land was ideal for corn and millet.

(   ) 5. Lots of villagers nicknamed this sort of land as "lazy land".

Development

III. Discuss the following questions.

1. What do you think of slash-and-burn?
2. Do you have more insights into slash-and-burn?

第二章 农耕思想
Chapter 2　Farming Ideology

Text A　天人合一价值观
Values of Unity of Heaven and Humans

Chinese nation is founded on agriculture since time immemorial and traditional agriculture has been thriving in China. Chinese farmers reclaim and plant; and Chinese farmlands are green in spring and yellow in autumn, contributing foodstuffs to Chinese people. Thus, a sea of Chinese people can live on the scant farmland in China. In the long-term farming practices, ancient Chinese people formed an idea of unity of heaven and humans which undergoes fermentation and evolution in the unique and vast soil in China, eventually crystallizing into a quintessential system of methodology or philosophy.

The Chinese character 天 (tian) is usually translated into "heaven" in English. But tian in Chinese language has more meanings. Originally it means the supreme being over the world. From this original meaning other meanings derived later: (a) the creator and supervisor of the world-god; (b) the origin of moral, political and social laws; (c) nature. Hence correspondingly in at least three senses we understand the idea of Unity of Heaven and Man.

Tian as the Supreme Being

In the first sense, tian was believed by ancient Chinese people to generate and supervise all men, and if anyone did good things he would be rewarded and he did bad things he would be punished. As tian could not come to the world to govern human beings in person, he sent his representatives-emperor or king. So in ancient China an emperor or a king was honored as "the son of heaven". As he was sent by god, the emperor must do his best to carry out the instruction of tian, that is to meet the people's needs: kept the society in order and made people live a happy life. Once he abused his power to do something against people's will, he would be warned and punished by heaven. If he continued to do wrong doings, he would be overthrown by his people whose will was regarded represent the will of heaven.

Such an idea is fully shown in one of Confucian classics *The Book of Documents*. It has the words of "Tian looks through the eyes of the people and listens through the ears of the people" and "what the people desire is always permitted by tian". It was also expressed by Dong Zhongshu, the famous philosopher in the Western Han Dynasty, in his book *Luxuriant Dew of the Spring and Autumn Annals*. He asserts that heaven and man are the same kind, and they communicate with each other and react to each other.

This conception influenced the mind of the Chinese people so deeply that rulers, leaders and governments of China in both ancient times and modern times must listen to the people, And just because of such a conception, people-oriented political idea was evolved, which became an important characteristic of traditional Chinese political culture.

Tian as the Origin of Divine Laws

In ancient China human beings' basic laws of morality, politics and society were regarded from heaven. These basic laws are called dao in Chinese language which originally means path, road or way. As these basic laws were regarded from heaven, they were taken as sacred and therefore should be observed by everybody. If someone failed to abide by these laws he would be regarded to violate the decree of heaven or run counter to the order of heaven, and thus he would be considered to lose his nature which was regarded to be endowed by heaven. Once he lost his nature given by heaven he lost the quality as a man and degenerated to an animal. On the contrary, if he could carry out these laws thoroughly and got great achievement in morality, he would be honored with the title of "sage" which was the noblest title in ancient China. If, so to speak, a king or an emperor was regarded as the political representative of heaven, then a sage should be taken as the moral representative of heaven, so a sage was also called "a king not in power" in the past.

Such an idea can be found in a Confucian classic called *The Doctrine of the Mean*, in which it said: "what heaven has conferred is called nature; being accordant with nature is called Dao; to learn and practice Dao is called education". Here Confucians connected social principles Dao with human nature and connected human nature with divine law—what heaven has conferred, in order to tell people that the three are essentially the same. That divine law, nature and Dao are finally boiled down to education means that moral education is essential and fundamental in social governance. So we are told in another Confucian doctrine *The Great Learning* that "from emperor down to commoners everybody must take moral cultivation as the foundation of life". Mencius also connected human being with heaven. He says: "Realizing wholly your conscience (which comprises, according to Mencius, the sense of sympathy, the sense of shamefulness, the sense of humbleness and the sense of right and wrong), you know your nature, and knowing your nature, you know heaven; realize your conscience and know your nature so that you can serve heaven." The words mean that as heaven endows us with human nature which shows us as conscience, if we achieve our whole conscience, we get a good understanding of the principles from heaven; and if we get a good understanding of the principles from heaven, we will be able to serve heaven (to do the most righteous things).

We find the explanatory words for Mencius' idea in the fore-mentioned book *The Doctrine of the Mean*. Only with extremist sincerity can man realize the nature from heaven and thereby realize the nature of human beings and the nature of things; with the realization of the nature of things, man can take part in the creation of things by heaven and earth; with partaking in the creation of things by heaven and earth man can be on a par with heaven and earth. What does it mean by "take part in"? The words from Dong Zhongshu, a philosopher in the Han Dynasty, may give a explanation. He says: "heaven, earth and man are the fundamentals of all the things; heaven produces them, earth breeds them and man achieves their values." Then how to achieve? Maybe we can refer it to Xun Tzu's opinion, who says in his book, "what a sovereign does is to harmonize the world; if his principles are just, all things will achieve their fitness, all live stocks their growth and all living beings their lives". By these words we find that in the minds of ancient Chinese people, man is not separated from heaven but an important assistant to it in that he helps to make a harmonious environment for all the things in the world.

Confucians highlighted moral education so much that their political philosophy could be summarized into rule of virtue. The sense it makes is that if all people cultivate themselves well or are educated to be virtuous, the whole society will be kept by people consciously and conscientiously in peace and order with few laws and penalties being carried out. Influenced by the idea of 'rule of virtue', an emperor, being 'the son and the representative of heaven' in the world, should cultivate himself to be a moral example while he asked his people to be virtuous.

Tian as Nature

As nature is also called Tian in China, the idea of Unity of Heaven and Man also means that man and his society is a part of nature and hence man must be in accordance with nature. Here Tian is not god, but it is regarded have virtues and laws which human beings should follow. Confucius had once said, "Only heaven can be regarded great, and only Yao can follow the example of it." For Confucians, heaven had the virtues of creativity (creates all things), activity (never stops) and modesty (never flaunts). In a Confucian classic, *The Book of Changes*, we can read words praising the virtues of heaven: "the greatest virtue of heaven and earth is creating", "heaven acts vigorous and people of virtue should follow it to make unceasing efforts in progress".

In the view of Taoists, the natural law was also our social and human law. Lao Tzu had once said, "man follows the example of earth, earth heaven, heaven Dao (the universal law), Dao natural laws." In Taoists' mind, the law that nature abides by is "doing nothing for every thing", which means nature never does a thing on purpose but achieves all the things in the world. According to this universal law, a ruler should do as little as possible in order not to interfere in the society if he wants his state prosperous, and a man should live as simply as possible if he wants a safe and long life. To do so, people should have as little as possible knowledge, desire, feeling and will.

Here we can see the great difference between Confucianism and Taoism. In the eye-mind of Confucians heaven (nature) is most active, diligent and prolific, so to follow the example of it people should undertake self-cultivation to become knowledgeable, able and moral, and thereby do their best in their duties and responsibilities for their family, nation and the whole world. On the contrary, Taoists regard nature as inactive, passive and undesirous, so to follow the example of it people should purify their minds, weaken their will and decrease their desires. In short, Confucians advocate a social life while Taoists pioneer a natural life.

Not only nature, some natural things can also be moral examples of Chinese people. For instance, Confucians praise jade for its purity, firmness, faithfulness and gentility while Taoists admire water as it is regarded humble and favorable.

By and large, the idea of Unity of Heaven and Man can be understood as "the divine is also the human, the social is also the natural". So unlike the western culture which makes qualitative differences between gods and man and nature and society, Chinese philosophy makes no clear boundary between divine beings and human beings and natural beings and social beings.

The idea of Unity of Heaven and Man has deeply and widely influenced Chinese culture and such an influence will continue in the future. So in Chinese people's mind god, nature and man coexist, affecting, relating to, reacting to and relying on each other, and as we live in such a great net of multiple relations, to make a peaceful world and lives a happy life we must make these relations harmonious.

The idea of unity of heaven and humans pushes Chinese farming thoughts and practices in acient times and gives birth to many farming theories such as, the theory of sancai and the theory of sanyi, putting emphasis on coordination of climate, geographic conditions and people. Macroscopically, under the guideline of the idea of unity of heaven and humans, the government would roll out a series of policies and measures to foster agricultural development and coordinate profits distribution. Microcosmically, under the guideline of the idea of unity of heaven and humans, the individual farmer would automatically tend to adopt farming model coordinating climate and geography, thus ensuring the sustainable development of agriculture.

Chinese agronomy in ancient times had been following the guideline of the idea of unity of heaven and humans and taking it into account in their farming practices. Admittedly, the idea of unity of heaven and humans which regards nature and society as an inseparable unity, coordinating agricultural production with nature conserve is undoubtedly perfectly correct in that it respects nature and highlights ecological development. Thus it is a glorious idea and is inherited and carried forward.

Words & Expressions

thrive/θraɪv/v. 茁壮成长，兴旺，繁荣
undergo/ˌʌndə'gəʊ/v. 经历，经受
fermentation/ˌfɜːmen'teɪʃn/n. 发酵
crystallize/'krɪstəlaɪz/vt. 使结晶；明确；使具体化；vi. 结晶，形成结晶
quintessential/ˌkwɪntɪ'senʃl/adj. 典型的，完美的；精髓的
derive/dɪ'raɪv/v. 获得，取得；起源于，来自；提取，衍生
correspondingly/ˌkɒrə'spɒndɪŋli/adv. 相应地，相对地
assert/ə'sɜːt/v. 坚称，断言；维护，坚持（权利或权威）；坚持自己的主张，表现坚定；
divine/dɪ'vaɪn/adj. 神的，天赐的；绝妙的，极令人愉快的；v. （凭直觉）猜测，推测；占卜，预测；n. [旧] 牧师，神学家；天意，上帝
sacred/'seɪkrɪd/adj. 神的，神圣的；宗教的，宗教性的；受尊重的，受崇敬的；不可侵犯的，不容干涉的
endow/ɪn'daʊ/v. 向（人，机构）捐赠，资助；赋予
degenerate/dɪ'dʒenəreɪt/v. 恶化，堕落，退化；adj. 衰退的，堕落的；退化的
thoroughly/'θʌrəli/adv. 完全地，极度地；仔细地，认真地，彻底地
morality/mə'ræləti/n. 道德，品行；正当性
sage/seɪdʒ/n. 圣人，贤人，哲人；鼠尾草；adj. 明智的；贤明的；审慎的
doctrine/'dɒktrɪn/n. 教义，主义，信条；（政府政策的）正式声明
confer/kən'fɜː(r)/v. 授予，赋予；商讨，交换意见
essentially/ɪ'senʃəli/adv. 本质上；本来
conscience/'kɒnʃəns/n. 良知，良心；内疚，愧疚
comprise/kəm'praɪz/v. 包括，包含；构成，组成
sympathy/'sɪmpəθi/n. 同情（心），理解；赞同，支持；（与某人的）同感，共鸣
breed/briːd/v. 交配繁殖；饲养，培育；养育，培养；引起，酿成；（通过核反应）增殖可裂变物质；n. 品种；（人的）类型，种类
sovereign/'sɒvrɪn/n. 君主，元首；金镑（旧时英国金币，面值一英镑）；独立国 adj. （国家）有主权的，完全独立的；掌握全部权力的，有至高无上的权力的；首要的；非常好的
harmonize/'hɑːmənaɪz/vt. 使和谐；使一致；以和声唱；vi. 协调；和谐；以和声唱
highlight/'haɪlaɪt/v. 突出，强调；用亮色突出；挑染；n. 最好（或最精彩、最激动人心）的部分；挑染的头发；强光部分
undertake/ˌʌndə'teɪk/v. 承担，从事；承诺，答应；[英] 在内侧行驶时赶上并超越（另一辆车）
coordination/kəʊˌɔːdɪ'neɪʃn/n. 协调，配合；身体的协调性；配位；同一等级（或类别）
macroscopically/ˌmækrəʊ'skɒpɪkli/adv. 肉眼可见地；宏观地；宏观地
microcosmically/ˌmaɪkrəʊ'kɒzmɪkli/adv. 从缩影角度；微观上，微观地
sustainable/sə'steɪnəb(ə)l/adj. （计划、方法、体制）可持续的，持续性的；（自然资

源）可持续的，不破坏环境的；站得住脚的

glorious/'glɔːriəs/adj. 光荣的，值得称道的；辉煌的，绚丽的；极其愉快的；（天气）阳光灿烂的

inherit/ɪn'herɪt/v. 继承（遗产）；经遗传获得（品质、身体特征等）；接手，承担；接收（前所有者的事物）；［古］（尤作圣经翻译和典故用语）得到

since time immemorial 自古以来

a sea of 大量；大量的

run counter to 与…相对；与…冲突

on the contrary 相反地；恰恰相反；正相反

boil down to 重点是；将…归结为

take part in 分担；参与

on a par with 与…同等

Notes

1. *The Supreme Being*：至上神，在我国，至上神主要围绕"帝"或"天"展开。从先秦至两汉，随着"帝"神圣性的消散和各类思想文化的融入，中国古代的至上神信仰呈现出以"天"为主要内涵、多种"天帝"形象为外延的状态。

2. *The Book of Documents*：《尚书》，最早书名为《书》，是一部追述古代事迹著作的汇编。分为《虞书》《夏书》《商书》《周书》。因是儒家五经之一，又称《书经》。通行的《十三经注疏》本《尚书》，就是《今文尚书》和伪《古文尚书》的合编本。现存版本中真伪参半。

3. *Luxuriant Dew of the Spring and Autumn Annals*：《春秋繁露》，是中国汉代哲学家董仲舒所作的政治哲学著作。《春秋繁露》推崇公羊学，发挥"春秋大一统"之旨，阐述了以阴阳五行、天人感应为核心的哲学-神学理论，宣扬"性三品"的人性论、"王道之三纲可求于天"的伦理思想及赤黑白三统循环的历史观，为汉代中央集权的封建统治制度奠定了理论基础。

4. *The Doctrine of the Mean*：《中庸》，是中国古代论述人生修养境界的一部道德哲学专著，是儒家经典之一，原属《礼记》第三十一篇，相传为战国时期子思所作。其内容肯定"中庸"是道德行为的最高标准，认为"至诚"则达到人生的最高境界，并提出"博学之，审问之，慎思之，明辨之，笃行之"的学习过程和认识方法。宋代学者将《中庸》从《礼记》中抽出，与《大学》《论语》《孟子》合称为"四书"。宋元以后，成为学校官定的教科书和科举考试的必读书，对中国古代教育和社会产生了极大的影响。其主要注本有程颢《中庸义》、程颐《中庸解义》、朱熹《中庸章句》、李塨《中庸传注》、戴震《中庸补注》、康有为《中庸注》、马其昶《中庸谊诂》和胡怀琛《中庸浅说》等。

5. *The Great Learning*：《大学》，是一篇论述儒家修身、齐家、治国、平天下思想的散文，原是《小戴礼记》第四十二篇，相传为春秋战国时期曾子所作，实为秦汉时儒家作品，是一部中国古代讨论教育理论的重要著作。

6. *Mencius*：孟子，名轲，字子舆（约公元前372—前289），邹国（今山东邹城东

南）人。战国时期哲学家、思想家、政治家、教育家，是孔子之后、荀子之前的儒家学派的代表人物，与孔子并称"孔孟"。

7. Xun Tzu：荀子（约公元前313—前238），名况，字卿（一说时人相尊而号为卿），战国末期赵国人，两汉时因避汉宣帝询名讳称"孙卿"，著名的思想家、哲学家、教育家，儒家学派的代表人物，先秦时代百家争鸣的集大成者。

8. Lao Tzu：老子，姓李，名耳，字聃，一字伯阳，或曰谥伯阳，春秋末期人，生卒年不详，籍贯也多有争议，《史记》等记载老子出生于楚国或陈国。中国古代思想家、哲学家、文学家和史学家，道家学派创始人和主要代表人物，与庄子并称"老庄"。后被道教尊为始祖，称"太上老君"。在唐朝，被追认为李姓始祖。曾被列为世界文化名人，世界百位历史名人之一。

9. The theory of Sancai：三才，姓名学术语，属于姓名学之五格剖象法术语。在五格剖象法中，三才为天格、人格、地格的总称。所谓三才，即天才、人才、地才，它们分别是天格、人格、地格数理的配置组合，反映综合内在运势。

10. The theory of sanyi：三易，夏代的《连山》、商代的《归藏》、周代的《周易》，并称为三易。东汉学者桓谭在《新论正经》中说："《连山》八万言，《归藏》四千三百言（秦朝精简本）。《连山》藏于兰台，《归藏》藏于太卜。"《连山》与《归藏》魏晋之后下落不明或被佛道吸收作经或亡佚，成为中华文化领域里的千古之谜。

Reading Comprehension

I. Each piece of the following information is given in one of the paragraphs in the passage. Identify the paragraph from which the information is derived and put the corresponding number in the space provided.

(   ) 1. Such an idea is fully shown in one of Confucian classics *The Book of Documents*.

(   ) 2. As Tian could not come to the world to govern human beings in person, he sent his representatives-emperor or king.

(   ) 3. In ancient China human beings' basic laws of morality, politics and society were regarded from heaven.

(   ) 4. In the view of Taoists, the natural law was also our social and human law.

(   ) 5. The Chinese character 天 (tian) is usually translated into 'heaven' in English.

II. Decide whether the statements are true (T) or false (F) according to the passage.

(   ) 1. But tian in Chinese language has more meanings than ten.

(   ) 2. So in ancient China an emperor or a king was honored as "heaven".

(   ) 3. Tian as the Supreme Being is fully shown in one of Mencius' classics *The Book of Documents*.

(   ) 4. Such an idea can be found in a Confucian classic called *The Doctrine of the Mean*.

(   ) 5. We find the explanatory words for Confucian this idea in the fore-mentioned book *The Doctrine of the Mean*.

Language Focus

III. Complete the sentences with the correct form of the words in the table.

assert	reward	degenerate	Confucius	Highlight
abide	fundamental	corresponding	conscience	Morality

1. If anyone did good things he would be _____ and he did bad things he would be punished.

2. In ancient China human beings' basic laws of _____ politics and society were regarded from heaven.

3. Such an idea can be found in a _____ classic called *The Doctrine of the Mean*.

4. He says: "heaven, earth and man are the _____ of all the things."

5. He _____ that heaven and man are the same kind, and they communicate with each other and react to each other.

6. Confucians _____ moral education so much that their political philosophy can be summarized into rule of virtue.

7. In Taoists' mind, the law that nature _____ by is "doing nothing for everything".

8. If we achieve our whole_____, we get a good understanding of the principles from heaven.

9. Once he lost his nature given by heaven he lost the quality as a man and _____ to an animal.

10. Hence _____ in at least three senses we understand the idea of Unity of Heaven and Man.

IV. Match the sentences in Section A with the English translation in Section B.

Section A

1．汉字"天"翻译成英文通常为"heaven"。
2．"民之所欲，天必从之"。
3．天地人，万物之本也，天生之，地养之，人成之。
4．人法地，地法天，天法道，道法自然。
5．不仅是大自然，自然性事物也能成为中国的榜样。

Section B

1. Lao Tzu had once said, "man follows the example of earth, earth heaven, heaven Dao (the universal law), Dao natural laws".

2. What the people desire is always permitted by tian.

3. Not only nature, some natural things can also be moral examples of Chinese people.

4. The Chinese character 天 (tian) is usually translated into "heaven" in English.

5. He says: "heaven, earth and man are the fundamentals of all the things; heaven produces them, earth breeds them and man achieves their values."

V. Translate the paragraph into Chinese.

The idea of unity of heaven and humans pushes Chinese farming thoughts and practices in ancient times and gives birth to many farming theories, such as, the theory of SanCai and the theory of SanYi, putting emphasis on coordination of climate, geographic conditions and people. Macroscopically, under the guideline of the idea of unity of heaven and humans, the government would roll out a series of policies and measures to foster agricultural development and coordinate profits distribution. Microcosmically, under the guideline of the idea of unity of heaven and humans, the individual farmer would automatically tend to adopt farming model coordinating climate and geography, thus ensuring the sustainable development of agriculture.

Chinese agronomy in ancient times had been following the guideline of the idea of unity of heaven and humans and taking it into account in their farming practices. Admittedly, the idea of unity of heaven and humans which regards nature and society as an inseparable unity, coordinating agricultural production with nature conserve is undoubtedly perfectly correct in that it respects nature and highlights ecological development. Thus it is a glorious idea and is inherited and carried forward.

Development

VI. Discuss the following questions.

1. What do you think are the influences of the idea of unity of heaven and humans on Chinese agriculture?

2. Do you have more insights into the influence of the idea of unity of heaven and humans on Chinese agriculture?

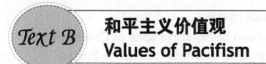

Text B 和平主义价值观 / Values of Pacifism

According to modern historical study, people living in China began an agricultural life at least 9000 years ago. Thereafter though there have been advanced handicraft industry and commercial economy in China, Chinese society remained to be agricultural until recent years. No society of other styles is more peace-loving than agricultural society: peasants live for generations in a village with their families, livestock and unmovable houses and lands, and wars will drive them out of the lands on which they live, destroy their houses, rob their properties and kill their families. So they do not like wars exerted on them, and also they do not like to wage wars on others unless they have to do so.

As a typical nation of agriculture for thousands of years, China has long evolved the value of peace-loving which is formally called pacifism. Such a value is embedded deeply in China's long history of agricultural social style and well shown in the main cultural streams throughout the history. Confucianists advocated the Society of Great Harmony which is described as "no conspiracy, no theft and no robbery would happen and hence gates need not to be shut at night". Thereby they opposed wars intensely. Confucius, according to what is recorded in *The Analects of Confucius*, had once been consulted by Duke Wei Ling for tactical deployment of troops, he replied, "about etiquette I have really learned something, but about military I've learned nothing." And the next day he left State Wei, showing his disagreement with the duke who loved war rather than peace. Mencius also says, "there was no just war recorded in *The Spring and Autumn Annals*", implying that he dislike wars. Taoists are more antagonistic than Confucianists to wars. In the book of *Tao Te Ching*, Lao Tzu says, "Those who serve a monarch with Dao will not use military force to seek hegemony in the world" and "Weapons are inauspicious and everything hates them, thereby people of Dao will not be close to them". To show his peace-loving values Lao Tzu even declares "requite evil with good". Even the great strategist Sun Tzu says in his work *The Arts of War*, "Military action is of vital importance to a state, concerning life and death, survival and downfall, hence it is a subject of careful inquiry". Therefore he advocates "subdue the opponent troops without fight". These, we can say, are the reflections of the traditional Chinese values of pacifism in military concepts.

The pacifism of traditional Chinese culture is also indicated by the popularity of Buddhism in China. Buddhism advocates "no killing" and "bosom of mercy" and hence it is regarded the most peace-loving religion in the world. Since the first century of Christian era foreign religions have begun their spreading in China, and only Buddhism quickly gained its popularity among Chinese people and even has become a powerful competitor to Chinese native religious once upon a time. Now Buddhism becomes even more popular in China and more and more Chinese people become Buddhism believers. This phenomenon reflects the pacifism in Chinese religious culture.

If we study the wars in the history of China we find that most of those between China and another country were defensive rather than offensive. The Great Wall winding thousands of miles among the mountains in the north of China was used for defending not attacking. To avoid international disputes and create a good international environment rulers of each dynasty in Chinese history would prefer marriage to military force. The most famous example of this is the marriage of Princess Wencheng and King Sontzen Gampo between the Tang Empire and the Kingdom Tubo in the 7th century. The marriage was taken as a peacemaker and Princess Wencheng was honored as a "peace envoy" in Chinese history. There still have been many other princesses in different periods of Chinese history who were married to neighboring states for peace, but here we cannot introduce them all.

The proverb "men of virtue solve disputes by mouth not fists" may be the most popular expression of traditional Chinese pacifism among Chinese masses.

Words & Expressions

thereafter/ˌðeərˈɑːftə(r)/*adv.* 其后；从那时以后
evolve/ɪˈvɒlv/*v.* 进化，演化；逐步发展，逐渐演变
embed/ɪmˈbed/*v.* （使）嵌入，把…插入；深信，使深留脑中；（计算机中）内置；栽种；派遣；*n.* 随军记者
conspiracy/kənˈspɪrəsi/*n.* 阴谋，密谋；阴谋集团
hence/hens/*adv.* 因此；之后
analects/ˈænəlekts/*n.* 文选
etiquette/ˈetɪkət/*n.* 礼节，规矩
subdue/səbˈdjuː/*vt.* 征服；抑制；减轻
bosom/ˈbʊzəm/*n.* 胸；胸怀；中间；胸襟；内心；乳房；内部；*vt.* 怀抱；把…藏在心中 *adj.* 知心的；亲密的
buddhism/ˈbʊdɪzəm/*n.* 佛教
envoy/ˈenvɔɪ/*n.* 使者，使节，（谈判等的）代表；公使（级别在大使之下）
according to/əˈkɔːdɪŋ tə/*prep.* 根据；据（…所说）；按（…所报道）；依照；按照
wage war on 设法强行控制，对…开战
gain popularity 受到欢迎

Notes

1. **The Society of Great Harmony**：大同社会是中国人思想传统中最后理想社会或人类社会的最高阶段。大同社会，是全民公有的社会制度，包括权力公有和财物公有，而首先是权力的公有。权力公有的口号是"天下为公"，具体措施是选贤与能，讲信修睦。

2. *The Analects of Confucius*：《论语》，是春秋时期思想家、教育家孔子的弟子及再传弟子记录孔子及其弟子言行而编成的语录文集，成书于战国前期。全书共 20 篇 492 章，以语录体为主，叙事体为辅，较为集中地体现了孔子及儒家学派的政治主张、伦理思想、道德观念及教育原则等。作品多为语录，但辞约义富，有些语句、篇章形象生动，其主要特点是语言简练，浅近易懂，而用意深远，有一种雍容和顺、纡徐含蓄的风格，能在简单的对话和行动中展示人物形象。

3. **Duke Wei Ling**：魏无忌（？—前243），即信陵君，魏国公子，与春申君黄歇、孟尝君田文、平原君赵胜并称为"战国四公子"。是战国时期魏国著名的军事家、政治

家，魏昭王少子、魏安厘王的异母弟。公元前 276 年，被封于信陵（河南宁陵县），所以后世皆称其为信陵君。

4. *The Spring and Autumn Annals*：《春秋》是我国古代史类文学作品。又称《春秋经》《麟经》或《麟史》等。《春秋》用于记事的语言极为简练，然而几乎每个句子都暗含褒贬之意，被后人称为"春秋笔法""微言大义"。它是中国古代儒家典籍"六经"之一，是我国第一部编年体史书，也是周朝时期鲁国的国史，现存版本据传是由孔子修订而成。

5. *Tao Te Ching*：道德经，《道德经》，春秋时期老子（李耳）的哲学作品，又称《道德真经》《老子》《五千言》《老子五千文》，是中国古代先秦诸子分家前的一部著作，是道家哲学思想的重要来源。道德经分上下两篇，原文上篇《德经》、下篇《道经》，不分章，后改为《道经》37 章在前，第 38 章之后为《德经》，并分为 81 章。

6. *Sun Tzu*：孙子，孙武（约公元前 545—前 470），字长卿，春秋末期齐国乐安（今山东省北部）人。中国春秋时期著名的军事家、政治家，尊称兵圣或孙子（孙武子），又称"兵家至圣"，被誉为"百世兵家之师""东方兵学的鼻祖"。

7. *The Arts of War*：《孙子兵法》，又称《孙武兵法》或《吴孙子兵法》，是中国现存最早的兵书，也是世界上最早的军事著作，早于克劳塞维茨《战争论》约 2300 年，被誉为"兵学圣典"。现存共有 6000 字左右，共十三篇。作者为春秋时祖籍齐国乐安的吴国将军孙武。

Reading Comprehension

I. Each piece of the following information is given in one of the paragraphs in the passage. Identify the paragraph from which the information is derived and put the corresponding number in the space provided.

(　　) 1. This phenomenon reflects the pacifism in Chinese religious culture.

(　　) 2. According to modern historical study people living in China began an agricultural life at least 9000 years ago.

(　　) 3. Such a value is embedded deeply in China's long history of agricultural social style and well shown in the main cultural streams throughout the history.

(　　) 4. The proverb "men of virtue solve disputes by mouth not fists" may be the most popular expression of traditional Chinese pacifism among Chinese masses.

(　　) 5. If we study the wars in the history of China we find that most of those between China and another country were defensive rather than offensive.

II. Decide whether the statements are true (T) or false (F) according to the passage.

(　　) 1. No society of other styles is more peace-loving than agricultural society.

(　　) 2. As a typical nation of agriculture for thousands of years, China has long evolved the value of peace-loving which is formally called pacifism.

(　　) 3. The pacifism of traditional Chinese culture is also indicated by the popularity of Taoism in China.

(　　) 4. The Great Wall winding thousands of miles among the mountains in the north of China was used for attacking.

(　　) 5. According to modern historical study people living in China began an agricultural life at least 10000 years ago.

III. Discuss the following questions.

1. What are the characteristics of the values of pacifism in China?
2. Do you have more insights into the values of pacifism in China?

Text C　家庭主义价值观　Values of Familism

We have mentioned above that China has long been a nation of agriculture and thereby the majority of Chinese population in the past lived in rural areas where a village often comprised only one clan, and the clan lived and farmed usually for many generations without migration. So superficially it seems that the basic social unit of old China is village, but essentially it is family. In cities it was the same case. Whether people were occupied in business, manufacture or government family was their stronghold or base camp. Some cases in which someone broke the family fence and devoted themselves thoroughly to the country could not change the whole condition. Even the national government and the whole politics was somewhat familial, thereby there has been the political concept of old China that a sate shares the same structure with a clan.

Such family-centered values have long been evolved in China, even before China as a real country took shape. Maybe it is a primitive value which can be universally found in early societies in the world, but only in China it was continually strengthened and lived out the whole history, the reason being possibly that China, as a typical agricultural county, remained unchanged throughout the history. The germ of such an idea can be traced back to more than three thousand years ago when the Shang Dynasty was established. We have mentioned in the foregoing parts that god worship and ancestor worship were both recorded in the Inscriptions on Bones or Tortoise Shells. Though god worship was also taken by royal or imperial courts, its popularity has never exceeded ancestor worship throughout Chinese history. So ancestor temples could be seen everywhere in families in ancient China while god temples were only built in some public or imperial places. After several cultural movements god worship has almost perished, but ancestor worship still remained popular in modern China. How and why Chinese people attach great importance to ancestor worship is another question, yet it really makes Chinese people pay much attention to family.

However, it is in the Spring and Autumn Period that familism was highlighted as a moral values, especially when Confucianism came into being as a school of thought. In *The Analects of Confucius* we can find that filial piety is greatly advocated by Confucius and his disciples. In *The Great Learning* we read, "Make yourself cultivated, then make your family regulated, then make your state rightly governed, then make the world peaceful and happy". These words set up a life goal and the steps to the goal for ancient Chinese people. Among the four steps, the second (regulating family) is a key link. Family is a place where a man is born and bred, and where he seeks help, warmth, consolation and happiness. Without family he cannot survive, so his first responsibility in life is for his family. He works hard not only for his current family but also for his family history. If he gets great achievement in fame, wealth or official rank, he is regarded to bring honor to his ancestors and he will be praised by his descendants in his family tree book. Conversely if he does wrong doings he is regarded to bring shame to his ancestors, and thereby he will be abounded by his descendants, being not recorded or mentioned in his family tree book. Even now if you ask a Chinese student why he or she works so hard, he or she would reply, "I want my parents to be pleased, to be happy, to lead a good life." So when Chinese people think of a man's value in his life they will see more what he has done for his family than for himself.

Familism, making people think that it is their duty to help each other within their families or clans, has played very important part in social development in old China, especially in the times when social security system did not exist and thus the young, the old, the disable, the orphan and the poor needed to be cared for. Even today it still contributes to social stability and harmony. But on the other hand, being over family-centered may be very harmful to social and national developments in that it may make people exceedingly care about family interests and thereby overlook social and national interests, and what's more even make people seek familial interests at the cost of social and national ones. It is in some way also harmful to individual development in that it may hinder people, especially the young, from being ambitious and enterprising for their future. In general, familism as a value, being generated in and fit to the old agricultural society, should be at least weakened, if not wholly abandoned, as China has now changed into an industrial and commercial country and thereby more and more people live more independently, individually and nationally.

Words & Expressions

superficially /ˌsuːpəˈfɪʃəli/ *adv.* 表面地；浅薄地

stronghold /ˈstrɒŋhəʊld/ *n.* 要塞；大本营；中心地

universally /ˌjuːnɪˈvɜːsəli/ *adv.* 普遍地；人人；到处

strengthen /ˈstreŋθn/ *v.* 使（情感、决心等）更强烈，使（关系）更加紧密；巩固（地位），加强（实力）；（货币）升值；增强（体质）；为…提供更有力的理由（证据）；（风力或水力）变强；使（法令或处罚）更有效；加强，加固（物体或结构）

establish /ɪˈstæblɪʃ/ *v.* 建立，设立；证实，确定；发现，找出；使被接受，使得到承认；使（故事的角色）真实

inscription/ɪn'skrɪpʃn/n. （书首页的）题词；（石头或金属上）刻写的文字，铭刻，碑文
tortoise/'tɔːtəs/n. 龟，[脊椎] 乌龟（等于 testudo）；迟缓的人
exceed/ɪk'siːd/v. 超过，超出；超越（限制）；优于，胜过
perish/'perɪʃ/v.（人或动物）惨死，猝死；湮灭，毁灭；（木头、橡胶、食物等）腐烂，腐败；[英，非正式] 经受酷寒
disciple/dɪ'saɪp(ə)l/n. 门徒，信徒；弟子
consolation/kɒnsə'leɪʃn/n. 安慰，起安慰作用的人（物）
descendant/dɪ'sendənt/n. 后裔，子孙；派生物，衍生物；adj. 下降的；祖传的
conversely/'kɒnvɜːsli/adv. 相反地
abound/ə'baʊnd/v. 大量存在，有许多；富于，充满
exceedingly/ɪk'siːdɪŋli/adv. 非常，极其；[古] 在很大程度上
hinder/'hɪndə(r)/v. 阻碍，妨碍；adj.（尤指身体部位）后面的
ambitious/æm'bɪʃəs/adj. 有抱负的，野心勃勃的；费劲的，艰巨的；热望的
enterprising/'entəpraɪzɪŋ/adj. 有事业心的，有进取心的；有魄力的，有胆量的
devote oneself to 专心从事（研究）；一味贪玩
attach importance to 重视
come into being 形成；开始存在
school of thought 学派；学派的观点；思想流派；有一家思想学校
filial piety 孝道；孝顺
family tree book 家谱；家谱图
social security system 社会安全体系；社会保险制度；社会保障体系；社会保障制度；制度

The Inscriptions on Bones：甲骨文，是中国的一种古老文字，又称"契文""甲骨卜辞""殷墟文字"或"龟甲兽骨文"。是我们能见到的最早的成熟汉字，主要指中国商朝晚期王室用于占卜记事而在龟甲或兽骨上契刻的文字，是中国及东亚已知最早的成体系的商代文字的一种载体。

Reading Comprehension

I. Each piece of the following information is given in one of the paragraphs in the passage. Identify the paragraph from which the information is derived and put the corresponding number in the space provided.

(   ) 1. In *The Analects of Confucius* we can find that filial piety is greatly advocated by Confucius and his disciples.

(   ) 2. Such family-centered values have long been evolved in China, even before China as a real country took shape.

(   ) 3. Without family he cannot survive, so his first responsibility in life is for his family.

(   ) 4. Even today it still contributes to social stability and harmony.

(   ) 5. So superficially it seems that the basic social unit of old China is village, but essentially it is family.

II. Decide whether the statements are true (T) or false (F) according to the passage.

(   ) 1. China has long been a nation of agriculture and thereby the majority of Chinese population in the past lived in urban areas where a village often comprised only one clan.

(   ) 2. Maybe it is a primitive value which can be universally found in early societies in the world, but only in China it was continually strengthened and lived out the whole history, the reason being possibly that China, as a typical industrial county, remained unchanged throughout the history.

(   ) 3. Family is a place where a man is born and bred, and where he seeks help, warmth, consolation and happiness.

(   ) 4. So when Chinese people think of a man's value in his life they will see more what he has done for his family than for himself.

(   ) 5. It is in some way also beneficial to individual development in that it may hinder people, especially the young, from being ambitious and enterprising for their future.

Development

III. Discuss the following questions.

1. What are the characteristics of the values of Familisim in China?
2. Do you have more insights into the values of Familisim in China?

第三章 传统农作物

Chapter 3　Traditional Cultivated Crops

Text A　五谷　Five Cereals

　　Rice grains were discovered at the Banpo, Hemudu and Majiabang sites, an indication that Chinese farming began with domestication and cultivation of grains. The Chinese Five Cereals (五谷, wugu) are a group of five farmed crops, mainly cereals or grains (except soybeans, which are legumes), which were all important in ancient China and regarded as sacred, or their cultivation was regarded as a sacred boon from a mythological or supernatural source: there are various lists specified. More generally, the term wugu is a synecdoche which refers to the whole or totality of all grains or cultivated staple agricultural crops by means of (nominally five) examples such as barley, beans, hemp, millet, oats, peas, rice, sesame, soy beans. In fact, an actual, specific list or limitation to five plant crops is not necessarily implied or expressed in this generic use of wugu. The use of wu or five predates modern botanical and mathematical notions and is better understood in the context of the philosophical and symbolic use of numbers in Chinese culture: the concept of Wu Xing (sometimes referred to as the five elements) often includes many corresponding lists of groups of five, such as colors, musical tones, and cardinal directions. The ancient Chinese term gu, meaning "crops/grain", predates modern taxonomy and botanical nomenclature. Rather than being a scientifically precise concept, the meaning of gu more generally relates to a tradition of Chinese domesticated and cultivated staple crops, as opposed to fruits, vegetables, herbs, or wild plant foods. However, historical literature includes various specific lists which have been recorded at various times. The concept of wugu is historically important in Chinese culture, especially because of the central role which cultivated crops and the resulting food products have had in an extensive tradition of agriculturally-based civilization.

The name, the Chinese Five Cereals or wugu, is sometimes translated as "five grains" "five sacred grains" or "five crops" as the word "gu", derived from "valley" in this ancient set phrase is less specific than narrower translations than "crop". Although not traditional in literary circles, the phrase might best be translated as "five (mainly millable) seeds" for greater biological accuracy, including medicinal applications of the optional group member hemp.

Mythological Wugu Cultivation and Cultures

The sense of sacrality regarding wugu proceeds from their being traditionally sourced from saintly rulers creating China's foundational culture as a whole. What is being claimed is not merely these culture givers preferring wugu among many crops, but that it is because of the wugu that agrarian society became possible, marking a profound, legendarily sudden shift from hunter-gatherer and nomadic patterns of life, which surrounding tribes continued to practice, thus making the central and lower Yellow River plains a distinctively settled heartland, a center or middle for those tribes which would come to consider themselves the central Middle Kingdom, or China. There are various traditions, often appearing in the same ancient Chinese texts such as the Shiji regarding which mythological early Emperor of China introduced the wugu, generally as part of the motif of the acquisition of agricultural civilization. It is widely believed that it was Shennong who cultivated baigu (literally means hundred types of grains in Chinese. Among these grains five of them (concisely named wugu as a whole) were staples of daily life for ancient Chinese. The mythico-political associations of the five grains can be seen as early as in the story of Boyi and Shuqi, who as a protest against the overthrow of the Shang Dynasty by the Zhou Dynasty ostentatiously refused to eat the five grains. Later the concept of refusing to eat the five grains underwent a complex development in the concept of bigu. The sacrality of the five grains is underlined by the belief that one of the sins which could lead one to suffer in the afterlife by being sent to Hell was by committing the sin of "squandering the five grains".

Archaeological Cultivation and Cultures

Archaeologically, Pengtoushan culture ca. 7500 BC—6100 BC on the Yangtze River has left tools of rice farming in some locations, though not at the first type site; Hemudu culture ca. 5000 BC—4500 BC by the coast South of the Yangtze River cultivated rice. Xinglongwa culture ca. 6200 BC—5400 BC in far Northeastern China around the Inner Mongolia-Liaoning border ate millet, apparently through agriculture. Dadiwan culture ca. 5800 BC—5400 BC, along the upper Yellow River ate millet; Nanzhuangtou culture ca. 8500 BC—7700 BC on the middle Yellow River around Hebei Province had grinding tools. Later Yangshao culture, ca. 5000 BC—3000 BC on the Yellow River grew millet extensively and sometimes barley and rice and vegetables, wove hemp and silk which depends on a supply of mulberry leaves and other silkworm fodder, but may have been limited to slash and burn methods.

Succeeding cultures on the Yellow River and elsewhere, which are indeed archaeologically dated with some rough connection to the legendary traditional dating of mythological agriculture-founding Three Sovereigns and Five Emperors reigns, which were early explicated by ancient Chinese historians as possible whole dynasties of reigns rather than single individuals, would increasingly suggest wider-scale and more intensive farming methods which facilitated the development of the rest of Chinese civilization by making ordinary daily provisioning a lesser concern, at least for some. By Longshan culture ca. 3000 BC—2000 BC along the Yellow River, sericulture, farming of domesticated silkworms for silk production is found, and definite cities appear, which depend on more advanced farm supply. Karuo culture ca. 3300 BC—2000 BC in Tibet, now Western China, cultivated foxtail millet.

Mythological and Historical Lists of Wugu Seeds and Crops

While wugu is a familiar word for most Chinese nowadays, it is far from being a well-defined concept. There are various versions of which five crops or even broad categories of crops are meant by this list.

The Chinese Five Cereals may be first listed in *The Book of Fan Shengzhi* on farming circa 2800 BC entitled *The Book of Fan Shengzhi*, which must be a mythological text as such a date predates coherent Chinese history texts as well Shang Dynasty oracle bone inscriptions.

The Classic of Rites compiled by Confucius (traditionally September 28th, 551 BC—479 BC) from older sources lists soybeans, wheat, proso millet, foxtail millet and hemp. Written records in the Han Dynasty (202 BC—220 AD) showed that wugu might be hemp, broomcorn, millet, wheat and beans. The other version includes rice, broomcorn, millet, wheat and beans. Some people believe that while hemp was edible, it was usually made into clothes and it should not be included in wugu, and that as the economic center shifted from the north to the south of China, where rice is the major grain, hemp was gradually replaced by rice. Thus the second version was more commonly recognized in China.

Another version replaces hemp with rice, whose cultivation is a preoccupation of texts throughout Chinese historical culture, and the first cultivation of rice is of continuing sociopolitical as well as anthropological interest in Chinese archaeology and that of other Asian regions.

Tang Dynasty (618—907) Buddhist Master Daoxuan's *Ritual of Measuring and Handling Light and Heavy Property* (《量处轻重仪》) lists 5 categories rather than 5 single species of seeds: house grain (fanggu, 房谷), loose grain (sangu, 散谷), horn grain (jiaogu, 角谷), beard grain (manggu, 芒谷), and cart grain (yugu, 与谷), and in species form these categories include foxtail millet (su, 粟), sorghum (shu, 黍, or gaoliang, 高粱), proso millet also known as broomcorn millet (shu, 菽, or ji, 稷), paddy, which implies rice (dao, 稻), wheat (mai, 麦), perilla (ren, 荏), all kinds of beans (dou), and linseed (ma or jusheng). Huilin's (慧琳) *Pronunciation and Meaning of All Scriptures* (《一切经音义》) cites Yang Chengtian's (阳承天) dictionary *Assembly of Characters* (《字统》), with the similar categories suigu, sangu, jiaogu, qigu, and shugu. Other versions including one modern Chinese dictionary note other candidates including sesame, barley, oats, and peas.

Despite the dispute over the precise definition of wugu, grains of these kinds were domesticated and cultivated over 7000 years ago in China. According to archaeological discoveries, tools for grinding grains were already invented by people living along the Yellow River in Hebei Province between 8500 BC and 7700 BC; millet was found to be widely cultivated in north China, especially in the Yellow River Valley, over 6000 years ago; carbonized rice grains were unearthed at some archaeological sites, like Hemudu in Zhejiang Province and Shanlonggang in Hunan Province, some of which date back to as early as 9000 years ago.

Besides wugu, ancient Chinese domesticated and cultivated many other grains, and some other grains such as corn were later introduced into China. However, wugu is sill popular in Chinese culture but it has become a joint name including all types of grains. No matter what grains they grow, people in rural China customarily hang a red banner onto their barn with four characters written on it—wugu fengdeng (literally means grain harvest). This is expected to bring a bumper grain harvest in the coming year and is part of the tradition for Chinese people to celebrate the lunar new year.

Domestication of Animals and Plants: Wugu and Liuchu

Besides grains, animals also began to be domesticated. Six kinds of animals, namely horse, buffalo, sheep, rooster, dog and pig, were already fed as livestock by the time the Neolithic Age finished.

The six kinds of animals were raised and trained due to needs of life and farming: horses and buffaloes were needed to pull ploughs; sheep provided milk for people to drink; dogs were kept to protect residents' settlements day and night; the rooster crowed to wake up people in the morning for a day's work; pigs produced meat as food.

These domesticated animals or liuchu (literally means the six kinds of livestock) in Chinese, fell into two classes: horses and buffaloes became helping hands for farming, with horses also serving as a means of transportation, and white sheep, which signified luck, were used as sacrifice for ancestors or gods. Therefore, the three animals were included in the positive class. In contrast, roosters were of less importance, dogs barked at strangers, which was thought to bully the weak, and pigs were lazy-eating and sleeping all the day. Thus, these three were classified as negative class.

During the lunar new year, families in rural areas will hang pieces of red papers with four characters liuchu xingwang, which means the growing number of healthy livestock, on the hen houses, cattle pens or doghouses. For Chinese farmers, wugu fengdeng together with liuchu xingwang means a well-off life.

Words & Expressions

legume/legju:m；lɪ'gju:m/*n*. 豆类；豆科植物；豆荚

boon/bu: n/*n*. 恩惠；福利；利益；*adj.* 愉快的；慷慨的

homophonic/ˌhɒməˈfɒnɪk/*adj.* 齐唱的；同音异义的
biological/ˌbaɪəˈlɒdʒɪkəl/*adj.* 生物的，生物学的，生命的；亲生的；含酶的
sacrality/seɪˈkrælətɪ/*n.* 神圣
merely/ˈmɪəlɪ/*adv.* 仅仅，只（用于强调某物小或微不足道）；无非是…
agrarian/əˈɡreərɪən/*adj.* 土地的；耕地的；有关土地的
profound/prəˈfaʊnd/*adj.* （影响）深刻的，极大的；*n.* 深海，海洋；深刻
extensively/ɪksˈtɛnsɪvlɪ/*adv.* 广阔地；广大地
explicate/ˈɛksplɪˌkeɪt/*vt.* 说明，解释
facilitate/fəˈsɪlɪˌteɪt/*v.* 使更容易，使便利；促进，推动
domesticate/dəˈmɛstɪˌkeɪt/*vt.* 驯养；教化；引进；*vi.* 驯养
Confucius/kənˈfjuːʃəs/*n.* 孔子（中国哲学家，教育家）
replace/rɪˈpleɪs/*v.* 取代；（用…）替换
anthropological/ˌænθrəpəˈlɒdʒɪkəl/*adj.* 人类学的；人类学上的
candidate/ˈkændɪˌdeɪt/*n.* 候选人，申请者
derive from 源出，来自，得自；衍生于
benefit from 得益于；得利于；因…而得到好处
proceed from 从…出发
as a whole 总的来说

Notes

1. Pengtoushan culture：彭头山文化，是中国的一种新石器时代文化，距今约 8300—9000 年，主要分布于长江流域，位于现湖南省北部，是中国史前文化的代表。

2. Xinglongwa culture：兴隆洼文化是分布于是内蒙古及东北地区的新石器时代文化，因首次发现于内蒙古自治区原敖汉旗宝国吐乡（现兴隆洼镇）兴隆洼村而得名，距今约 8000 年，经济形态除农耕外兼狩猎、采集。

3. Nanzhuangtou culture：南庄头文化，南庄头遗址，位于河北省保定市徐水区南庄头村东北约 2 千米处，萍河与鸡爪河之间。泥河湾、周口店、山顶洞等古文化遗址环绕着它，是中国北方地区年代最早的新石器时代的遗址，距今 10500 至 9700 年左右。

4. Three Sovereigns and Five Emperors：三皇五帝，是"三皇"与"五帝"的合称。原始意义上的三皇是指远古三皇（天皇氏、地皇氏、人皇氏），后增补伏羲氏等作为"三皇"。原始意义上的五帝是指远古五方上帝，后增补黄帝公孙轩辕（也称姬轩辕）等五位上古部落首领作为"五帝"。

5. Longshan culture：龙山文化，泛指中国黄河中、下游地区约新石器时代晚期的一类文化遗存，属铜石并用时代文化。

6. Karuo culture：卡若文化遗址，距离西藏昌都镇以西约 12 千米处。卡若文化遗址，被发现有大量的石斧、石锛和陶罐等原始文物和工具。发现房屋遗迹 31 处，出土石器、骨器和陶片等，以及大量的动物骨骼和粟米，经测定这些物品均出自 4000 到 5000 年前的"新石器时期"，对研究藏文化的发展尤其是西藏早期历史和汉藏关系史有重要意义。

7. *The Classic of Rites*：《礼记》，又名《小戴礼记》《小戴记》，成书于汉代，为西汉礼学家戴圣所编。《礼记》是中国古代一部重要的典章制度选集，共 20 卷 49 篇，书中内容主要写先秦的礼制，体现了先秦儒家的哲学思想（如天道观、宇宙观、人生观）、教育思想（如个人修身、教育制度、教学方法、学校管理）、政治思想（如以教化政、大同社会、礼制与刑律）、美学思想（如物动心感说、礼乐中和说），是研究先秦社会的重要资料，是一部儒家思想的资料汇编。

8. *Master Daoxuan*：道宣（596—667），俗姓钱，字法遍，原籍吴兴长城（今浙江长兴），一作丹徒人，自称吴兴人，生于京兆长安。唐代高僧，佛教南山律宗开山之祖，又称南山律师、南山大师，世称"律祖"。东汉富春侯钱让之后。

9. *Ritual of Measuring and Handling Light and Heavy Property*：《量处轻重仪》，佛教律仪书，二卷。唐道宣撰于 637 年（唐贞观十一年），667 年（乾封二年）重修。

10. Huilin's（慧琳）*Pronunciation and Meaning of All Scriptures*（《一切经音义》）：解释佛经字义的书。唐代大慈恩寺翻经沙门释玄应（释慧琳）撰，凡 25 卷。《唐书·艺文志》著录，称《众经音义》，当为原来名称。

11. Yang Chengtian's（阳承天）*dictionary Assembly of Characters*（《字统》）：唐朝和尚，阳承天撰写的字典类图书，《字统》。

Reading Comprehension

I. Each piece of the following information is given in one of the paragraphs in the passage. Identify the paragraph from which the information is derived and put the corresponding number in the space provided.

(　　) 1. What is being claimed is not merely these culture givers preferring wugu among many crops, but that it is because of the wugu that agrarian society became possible.

(　　) 2. By Longshan culture ca. 3000 BC—2000 BC along the Yellow River, sericulture, farming of domesticated silkworms for silk production is found, and definite cities appear, which depend on more advanced farm supply.

(　　) 3. Xinglongwa culture ca. 6200 BC—5400 BC in far Northeastern China around the Inner Mongolia-Liaoning border ate millet, apparently through agriculture.

(　　) 4. While wugu is a familiar word for most Chinese nowadays, it is far from being a well-defined concept.

(　　) 5. The name is sometimes translated as "five grains" "five sacred grains", or "five crops" as the word "gu", derived from "valley" in this ancient set phrase is less specific than narrower translations than "crop".

II. Decide whether the statements are true (T) or false (F) according to the passage.

(　　) 1. The Chinese Five Cereals are a group of 5 farmed crops, mainly cereals or grains (except soybeans, which are legumes).

(　　) 2. The sense of sacrality regarding wugu proceeds from their being traditionally sourced from saintly rulers creating China's foundational culture as a whole.

(　　) 3. Later Yangshao culture, ca. 5000 BC—3000 BC on the Yellow River grew wheat extensively and sometimes barley and rice and vegetables, wove hemp and silk which depends on a supply of mulberry leaves and other silkworm fodder, but may have been limited to slash and burn methods.

(　　) 4. Karuo culture ca. 3300 BC—2000 BC in Taiwan, now Western China, cultivated foxtail millet.

(　　) 5. The Classic of Rites compiled by Mencius (traditionally 551 BC—479 BC) from older sources lists soybeans, wheat, proso millet, foxtail millet and hemp.

Language Focus

III. Complete the sentences with the correct form of the words in the table.

archaeological	sacrality	domesticate	Confucius	candidate
biological	extensive	mere	regard	domesticate

1. The Chinese Five Cereals are a group of 5 farmed crops, mainly cereals or grains (except soybeans, which are legumes), which were all important in ancient China and regarded as sacred, or their cultivation was _____ as a sacred boon from a mythological or supernatural source.

2. The name, the Chinese Five Cereals or wugu, is sometimes translated as "five grains", "five sacred grains", or "five crops" as the word "gu", derived from "valley" in this ancient set phrase is less specific than narrower translations than "crop". Although not traditional in literary circles, the phrase might best be translated as "five (mainly millable) seeds" for greater _____ accuracy, including medicinal applications of the optional group member hemp.

3. What is being claimed is not _____ these culture givers preferring wugu among many crops, but that it is because of the wugu that agrarian society became possible, marking a profound, legendarily sudden shift from hunter-gatherer and nomadic patterns of life.

4. The sense of _____ regarding wugu proceeds from their being traditionally sourced from saintly rulers creating China's foundational culture as a whole.

5. Later Yangshao culture, ca. 5000 BC—3000 BC on the Yellow River grew millet _____ and sometimes barley and rice and vegetables, wove hemp and silk which depends on a supply of mulberry leaves and other silkworm fodder, but may have been limited to slash and burn methods.

6. Succeeding cultures on the Yellow River and elsewhere, which are indeed _____ dated with some rough connection to the legendary traditional dating of mythological agriculture thus it can be seen that the modern Chinese tea culture had basically taken _____ in the Ming Dynasty.

7. By Longshan culture ca. 3000 BC—2000 BC along the Yellow River, sericulture,

farming of _____ silkworms for silk production is found.

8. The Classic of Rites compiled by_____ (traditionally 551 BC—479 BC) from older sources lists soybeans, wheat, proso millet, foxtail millet and hemp.

9. Despite the dispute over the precise definition of wugu, grains of these kinds were _____ and cultivated over 7000 years ago in China.

10. Other versions including one modern Chinese dictionary note other _____ including sesame, barley, oats, and peas.

Ⅳ. Match the sentences in Section A with the English translation in Section B.

Section A

1. 根据考古发现，河北黄河流域居民在公元前 8500 至 7700 年就已经发明了谷类研磨工具。

2. 除了五谷，古代中国农民还驯化和栽培了许多其他的谷类作物，例如，玉米后来就被中国引进进行种植。

3. 中国文化学者之所以将其命名为"五谷"，不是因为他们更加偏爱这五种谷类作物，而是因为它们是农业社会形成的基础，标志了从采集狩猎和游牧生活向农业社会的重要历史性过渡。

4. 大约公元前 6200—前 5400 年，兴龙湾文化发源于中国东北的内蒙古和辽宁交界处，当时的居民已经开始食用粟类植物，证明他们已经会农业种植粟类作物了。

5. 中国"五谷"一词是五种农作物的统称，主要是谷类作物，不包括豆类的黄豆。

Section B

1. The Chinese Five Cereals are a group of 5 farmed crops, mainly cereals or grains (except soybeans, which are legumes).

2. What is being claimed is not merely these culture givers preferring wugu among many crops, but that it is because of the wugu that agrarian society became possible, marking a profound, legendarily sudden shift from hunter-gatherer and nomadic patterns of life.

3. Xinglongwa culture ca. 6200 BC—5400 BC in far Northeastern China around the Inner Mongolia-Liaoning border ate millet, apparently through agriculture.

4. According to archaeological discoveries, tools for grinding grains were already invented by people living along the Yellow River in Hebei Province between 8500 and 7700 BC.

5. Besides wugu, ancient Chinese domesticated and cultivated many other grains, and some other grains such as corn were later introduced into China.

Ⅴ. Translate the paragraph into Chinese.

Despite the dispute over the precise definition of wugu, grains of these kinds were domesticated and cultivated over 7000 years ago in China. According to archaeological discoveries, tools for grinding grains were already invented by people living along the Yellow River in Hebei Province between 8500 BC and 7700 BC; millet was found to be widely cultivated in north China, especially in the Yellow River Valley, over 6000 years ago; carbonized rice grains were unearthed at some archaeological sites, like Hemudu in Zhejiang Province and Shanlonggang in Hunan Province, some of which date back to as early as 9000 years ago.

VI. Discuss the following questions.

1. What are the roles of wugu in Chinese farming civilization?
2. Do you have more insights into wugu culture and expound them respectively?

中药
Traditional Chinese Medicine

Humanity has created a colorful global civilization in the long course of its development, and the civilization of China is an important component of the world civilization harboring great diversity. As a representative feature of Chinese civilization, traditional Chinese medicine (TCM) is a medical science that was formed and developed in the daily life of the people and in the process of their fight against diseases over thousands of years. It has made a great contribution to the nation's procreation and the country's prosperity, in addition to producing a positive impact on the progress of human civilization.

TCM has created unique views on life, on fitness, on diseases and on the prevention and treatment of diseases during its long history of absorption and innovation. It represents a combination of natural sciences and humanities, embracing profound philosophical ideas of the Chinese nation. As ideas on fitness and medical models change and evolve, traditional Chinese medicine has come to underline a more and more profound value.

Since the founding of the People's Republic of China in 1949, the Chinese government has set great store by TCM and rendered vigorous support to its development. TCM and Western medicine have their different strengths. They work together in China to protect people from diseases and improve public health. This has turned out to be one of the important features and notable strengths of medicine with Chinese characteristics.

In remote antiquity, the ancestors of the Chinese nation chanced to find that some creatures and plants could serve as remedies for certain ailments and pains, and came to gradually master their application. As time went by, people began to actively seek out such remedies and methods for preventing and treating diseases. Sayings like "Shennong (Celestial Farmer) tasting a hundred herbs" and "food and medicine coming from the same source" are characteristic of those years.

The discovery of alcohol in the Xia Dynasty (2070 BC—1600 BC) and the invention of herbal decoction in the Shang Dynasty (1600 BC—1046 BC) rendered medicines more effective.

In the Western Zhou Dynasty (1046 BC—771 BC), doctors began to be classified into four categories — dietician, physician, doctor of decoctions and veterinarian.

During the Spring and Autumn and Warring States Period (770 BC—221 BC), Bian Que drew on the experience of his predecessors and put forward the four diagnostic methods — inspection, auscultation & olfaction, inquiry, and palpation, laying the foundation for TCM diagnosis and treatment.

The Huangdi Neijing (*Yellow Emperor's Inner Canon*) compiled during the Qin and Han times (221 BC—220 BC) offered systematic discourses on human physiology, on pathology, on the symptoms of illness, on preventative treatment, and on the principles and methods of treatment. This book defined the framework of TCM, thus serving as a landmark in TCM's development and symbolizing the transformation from the accumulation of clinical experience to the systematic summation of theories. A theoretical framework for TCM had been in place.

The Shanghan Zabing Lun (*Treatise on Febrile Diseases and Miscellaneous Illnesses*) collated by Zhang Zhongjing in the Eastern Han Dynasty advanced the principles and methods to treat febrile diseases due to exogenous factors (including pestilences). It expounds on the rules and principles of differentiating the patterns of miscellaneous illnesses caused by internal ailments, including their prevention, pathology, symptoms, therapies, and treatment. It establishes the theory and methodology for syndrome pattern diagnosis and treatment differentiation. The *Shennong Bencao Jing* (*Shennong's Classic of Materia Medica*) — another masterpiece of medical literature appeared during this period — outlines the theory of the compatibility of medicinal ingredients. For example, it holds that a prescription should include at the same time the jun (or sovereign), chen (or minister), zuo (or assistant) and shi (or messenger) ingredient drugs, and should give expression to the harmony of the seven emotions as well as the properties of drugs known as "four natures" and "five flavors". All this provides guidance to the production of TCM prescriptions, safe application of TCM drugs and enhancement of the therapeutic effects, thus laying the foundation for the formation and development of TCM pharmaceutical theory. In the late years of the Eastern Han Dynasty, Hua Tuo (140—208) was recorded to be the first person to use anesthetic (mafeisan) during surgery.

The Zhenjiu Jiayi Jing (*AB Canon of Acupuncture and Moxibustion*) by Huangfu Mi during the Western Jin time expounded on the concepts of zangfu (internal organs) and jingluo (meridians and collaterals). This was the point when theory of jingluo and acupuncture & moxibustion began to take shape.

Sun Simiao, a great doctor of the Tang Dynasty (618—907), proposed that mastership of medicine lies in proficient medical skills and lofty medical ethics, which eventually became the embodiment of a moral value of the Chinese nation, a core value that has been conscientiously upheld by the TCM circles.

A herbology and nature masterpiece, the *Bencao Gangmu* (Compendium of Materia Medica) compiled by Li Shizhen in the Ming Dynasty was the first book in the world that scientifically categorized medicinal herbs. It was a pioneering work that advanced TCM pharmaceutical theory.

The Wenre Lun (*A Treatise on Epidemic Febrile Diseases*) by Ye Tianshi during the Qing Dynasty developed the principles and methods for prevention and treatment of pestilential febrile diseases. It represents the theory and results of the practice of TCM in preventing and treating such diseases.

Following the spread of Western medicine in China from the mid-Qing Dynasty, especially during the period of the Republic of China, some TCM experts began to explore ways to absorb the essence of Western medicine for a combination of TCM with Western medicine.

Words & Expressions

harbor/'hɑːbə/*n*. 港，海港；避难所
innovation/ˌɪnə'veɪʃən/*n*. 新事物，新方法；革新，创新
embrace/ɪm'breɪs/*v*. 拥抱；欣然接受，乐意采纳；*n*. 拥抱；接受，信奉
underline/ˌʌndə'laɪn/*v*. 在…下面画线；强调，突出
vigorous/'vɪɡərəs/*adj*. 精力充沛的，充满活力的；强壮的，强健的
remedy/'rɛmɪdɪ/*n*. 补救；治疗；药品
ailment/'eɪlmənt/*n*. 小病；不安
decoction/dɪ'kɒkʃən/*n*. 煎煮；煎熬的药；煮出的汁
render/'rɛndə/*v*. 使成为，使处于某种状态；给予，提供；（以某种方式）表达，表现
veterinarian/ˌvetərɪ'neərɪən/*n*. 兽医
aluscultation/ˌɔːskəl'teɪʃən/*n*. [临床] 听诊
olfaction/ɒl'fækʃən/*n*. 嗅觉
physiology/ˌfɪzɪ'ɒlədʒɪ/*n*. 生理学；生理机能
pathology/pə'θɒlədʒɪ/*n*. 病理学；病状；精神异常；社会异常；语言异常
landmark/'lændˌmɑːk/*n*. 陆标，地标；里程碑，转折点
exogenous/ɛk'sɒdʒɪnəs/*adj*. 外生的；外因的；外成的
pestilences/'pestɪləns/*n*. 瘟疫（尤指鼠疫）；有害的事物
expound/ɪk'spaʊnd/*v*. 详述，阐述；说明，讲解；解释…的意思
compatibility/kəmˌpætə'bɪlətɪ/*n*. 共存；和睦相处；（计算机设备的）兼容性
ingredient/ɪn'ɡriːdɪənt/*n*. （食品的）成分，原料；要素，因素
property/'prɒpətɪ/*n*. 所有物，财产；地产，房地产
plharmaceutical/ˌfɑːmə'sjuːtɪkəl/*adj*. 制药的；*n*. 药物
canon/'kænən/*n*. 原则；（主教座堂的）咏礼司铎；准则；标准；（某作家的）真作，精品

acupuncture/'ækjuˌpʌŋktʃə/n. 针灸，针刺疗法
moxibustion/ˌmɒksɪ'bʌstʃən/n. 艾灸
meridian/mə'rɪdɪən/n. 子午线，经线；（针灸及中医用语）经脉；最高点，顶点
collateral/kɒ'lætərəl/n. 抵押品，担保品；旁系亲属；adj. 附属的；旁系的；并行的
febrile/'fiːbraɪl/adj. 发热的；[医] 热病的；发烧的
set great store by 注重
take shape v. 形成；成形；体现；具体化

Notes

1. *The Huangdi Neijing (Yellow Emperor's Inner Canon)*：《黄帝内经》分《灵枢》《素问》两部分，是中国最早的医学典籍，传统医学四大经典著作之一（其余三者为《难经》《伤寒杂病论》《神农本草经》）。

2. *The Shanghan Zabing Lun (Treatise on Febrile Diseases and Miscellaneous Illnesses)*：《伤寒杂病论》是中国传统医学著作之一，是一部论述外感病与内科杂病为主要内容的医学典籍，作者是东汉末年张仲景，是中国中医院校开设的主要基础课程之一。《伤寒杂病论》系统地分析了伤寒的原因、症状、发展阶段和处理方法，创造性地确立了对伤寒病的"六经分类"的辨证施治原则，奠定了理、法、方、药的理论基础。

3. *The Zhenjiu Jiayi Jing (AB Canon of Acupuncture and Moxibustion)*：《针灸甲乙经》，又称《黄帝甲乙经》《黄帝三部针经》《黄帝针灸甲乙经》。西晋·皇甫谧撰，12卷，128篇，成书于公元282年。前六卷论述基础理论，后六卷记录各种疾病的临床治疗，包括病因、病机、症状、诊断、取穴、治法和预后等。采用分部和按经分类法，厘定了腧穴，详述了各部穴位的适应证和禁忌、针刺深度与灸的壮数，是我国现存最早的一部理论联系实际的针灸学专著。

4. *Bencao Gangmu (Compendium of Materia Medica)*：《本草纲目》，本草著作，52卷。明代李时珍（东璧）撰于1552年（嘉靖三十一年）至1578年（万历六年），稿凡三易。此书采用"目随纲举"编写体例，故以"纲目"名书。

5. *Wenre Lun (A Treatise on Epidemic Febrile Diseases)*：《温热论》为温病通论著作，由清代叶桂（天士）口述，其弟子顾景文执笔著录，全文十分简短，仅4000余字。为温病学说的奠基性著作。叶氏在此书中提出了温热病的卫气营血辨证体系，并论述了卫气营血相应的治法。此外，叶氏对通过观察舌象、皮肤斑疹及验齿来诊断温病的病情及判断预后方面也有独到的论述。

Reading Comprehension

I. Each piece of the following information is given in one of the paragraphs in the passage. Identify the paragraph from which the information is derived and put the corresponding number in the space provided.

(   ) 1. It was a pioneering work that advanced TCM pharmaceutical theory.

(   ) 2. They work together in China to protect people from diseases and improve public health.

(　　) 3. In the late years of the Eastern Han Dynasty, Hua Tuo was recorded to be the first person to use anesthetic (mafeisan) during surgery.

(　　) 4. TCM has created unique views on life, on fitness, on diseases and on the prevention and treatment of diseases during its long history of absorption and innovation.

(　　) 5. A theoretical framework for TCM had been in place.

II. Decide whether the statements are true (T) or false (F) according to the passage.

(　　) 1. Sayings like "Shennong (Celestial Farmer) tasting a thousand herbs" and "food and medicine coming from the same source" are characteristic of those years.

(　　) 2. In the Western Han Dynasty, doctors began to be classified into four categories — dietician, physician, doctor of decoctions and veterinarian.

(　　) 3. During the Spring and Autumn and Warring States period, Hua Tuo drew on the experience of his predecessors and put forward the four diagnostic methods — inspection, auscultation & olfaction, inquiry, and palpation, laying the foundation for TCM diagnosis and treatment.

(　　) 4. *Huang Di Nei Jing* (*Yellow Emperor's Inner Canon*) compiled during the Qin and Han times expounds on the rules and principles of differentiating the patterns of miscellaneous illnesses caused by internal ailments, including their prevention, pathology, symptoms, therapies, and treatment.

(　　) 5. Sun Simiao, a great doctor of the Song Dynasty, proposed that mastership of medicine lies in proficient medical skills and lofty medical ethics, which eventually became the embodiment of a moral value of the Chinese nation, a core value that has been conscientiously upheld by the TCM circles.

III. Discuss the following questions.

1. Can you find more information about Bian Que?
2. Can you find more information about the Huang Di Nei Jing?

园艺
Traditional Horticulture

History of Chinese Horticulture

The ancient history of Chinese horticulture began along the fertile plains of the Yellow River around 2000 BC and the writings of Confucius (552 BC—479 BC) are the first source to mention horticulture. Some crops grown back then were peaches (Prunuspersica), plums (Prunus salicina), Japanese apricot (Prunus mume), jujube, chestnut, mulberry, quince, Chinese cabbage, bottle gourds, and various melons. The Chinese developed cultivation techniques like row planting to space out plant root zones. Tools such as wheelbarrows, pottery, and hoes were utilized. The creation of irrigation canals allowed farmers to produce healthy crops in larger numbers. Eventually the Chinese began to fertilize their crops with "night soil" (human feces). They had an organic system where everything was composted, including bones, leather, manure, and straw.

Between 221 and 550 ornamental horticulture became a part of Chinese culture and spread throughout Asia with the development of emperor's rural and urban gardens. Biological control of insects was used in the culture of orange and litchi. The appearance of flower markets, flower arranging, bonsai and a great number of ornamental varieties of many plants allowed the ornamental horticulture of china to flourish in the first century.

The beginning of greenhouse growing structures in China dates back to the Han Dynasty when alliums were grown in heated structures during the winter. Later on during the Tang Dynasty (618—907), natural hot springs were used for forcing vegetables in the winter. Simple greenhouses were made using translucent oiled paper as covers to grow vegetables and flowers during the Song Dynasty. By the late 19th century, glasshouses were imported but these proved to be too expensive in terms of capitalization costs and heating. In the second half of the 20th century the use of fertilizers, pesticides, tractors, protected horticulture, and advances in genetic improvement were acquired from western countries. Windbreaks and solar lean-to greenhouses with polyvinyl chloride or polyethylene surfaces were built near Beijing and Shanghai. These structures shielded crops from the elements while transparent plastic surfaces provided the crops with ample light transmission.

Modern greenhouses from foreign countries began to appear around China by the late 20th century, they were made of glass, polycarbonate, or polyethylene, and had steel frames. These new greenhouses utilized advanced computerized control systems with automatic irrigation and environmental controls. These houses were too expensive and complicated for many Chinese growers. Low-cost, low-input plastic tunnels and solar lean-to houses experienced a sharp rise in popularity.

Today plastic high-tunnels and low-tunnels are widely used for commercial horticultural crop production all over China. In the north, northeast, and northwest, where winter is cold, plastic tunnels are mainly used in early spring and late autumn. In the south where winter is mild, plastic tunnels can be used all year round. Growers raised their own vegetable plants from seed, which took a long time, so in 1987 China established the first mechanized vegetable seedling production farm to produce mass quantities of plug-seedlings for peppers, cabbage, and cauliflower. This process allowed growers to grow a lot more crops, a lot faster.

Bonsai, one of classic showcase of Chinese horticulture

As one of most notable things in Chinese horticulture, Chinese Bonsai is the art of designing a miniature tree in a shallow pot or container. Bonsai (which translates to "tree in a pot") is also known as "pun-sai" and "penjing" (the word "pen" means container or pot and the word "jing" means scenery). Chinese Bonsai, inspired by nature, originated in China around 1300 years ago. Originally Bonsai was practiced only by the elite of ancient China. The miniature trees were considered a luxury and were given as gifts. Around 1100 Buddhist monks brought the Bonsai to Japan and the art was adopted by the Japanese. It was not until the early 1900's that Bonsai spread to the rest of the world.

Bonsai is often viewed as an object for meditation. The designing of the Bonsai tree is contemplative, a Zen practice. Chinese style meditation encourages liberation of the mind; encouraging it to flow in its own natural way. Creating Bonsai, arranging rocks in the miniature landscape, clipping and the adding of new elements is a process of active meditation. Something new may be discovered, and having flowed naturally out of the mind it is harmonious and lifts the spirit.

The purpose of Bonsai is not just to re-create nature in a pot, but to actually capture the spirit. The Chinese see the universe as having two sides of cosmic energy; this is called the yin and the yang. In a Bonsai tree, this is depicted through drama, rhythm and balance. Overall unity is important, therefore, many considerations have to be made. These include the type of container, the placement of the Bonsai tree, the species of the tree, the size, shape and color of the tree as well as other details such as the rocks which also has to be chosen just so.

Words & Expressions

horticulture/ˈhɔːtɪˌkʌltʃə/ n. 园艺，园艺学
utilize/ˈjuːtɪˌlaɪz/ v. 利用，使用
eventually/ɪˈventʃuəlɪ/ adv. 最终，结果
organic/ɔːˈgænɪk/ adj. 有机的；n. 有机物质；有机食品
compost/ˈkɒmpɒst/ n. 堆肥；混合物；vt. 堆肥；施堆肥
ornamental/ˌɔːnəˈmentəl/ adj. 装饰性的，作点缀的
biological/ˌbaɪəˈlɒdʒɪkəl/ adj. 生物的，生物学的，生命的；亲生的
flourish/ˈflʌrɪʃ/ v. 繁荣，昌盛；n. 夸张动作；（讲话或文章的）华丽辞藻
acquire/əˈkwaɪə/ v. 获得，得到；学到，习得；患上（疾病）；逐渐具有，开始学会
windbreak/ˈwɪndbreɪk/ n. 防风林；防风墙；防风物

polyvinyl chloride/ˌpɒlɪ'vaɪnɪl 'klɔːraɪd/*n*. [高分子]聚氯乙烯
polyethylene/ˌpɑːli'eθəliːn/*n*. 聚乙烯
polycarbonate/ˌpɒli'kɑːbənət/*n*. [高分子]聚碳酸酯
complicate/'kɒmplɪˌkeɪt/*v*. 使复杂化，使难以理解；引起并发症；使卷入，使陷入
cauliflower/'kɒlɪˌflaʊə/*n*. 花椰菜，菜花
notable/'nəʊtəbəl/*adj*. 显要的，值得注意的；非常成功的；*n*. 显要人物，名流
miniature/'mɪnɪtʃə/*adj*. 微型的，小型的；*n*. 缩微模型，微型复制品；微型画
inspire/ɪn'spaɪə/*v*. 激励，鼓舞；赋予灵感，激发（想法）
meditation/ˌmedɪ'teɪʃən/*n*. 冥想；沉思，深思；静坐
contemplative/'kɒntemˌpleɪtɪv/*adj*. 沉思的，冥思的；（对宗教问题）冥想的
arrange/ə'reɪndʒ/*v*. 安排，筹备；整理，布置，排列；改编；商定
capture/'kæptʃə/*v*. 俘获，捕获；夺取，占领；吸引；*n*. 捕获，被捕获
overall/ˌəʊvər'ɔːl/*adj*. 总的，全面的；所有的，包括一切的；*adv*. 全部，总共
space out 使间隔开；（因服食毒品等）变得昏昏沉沉；加宽行距或字距
in terms of 依据；按照；在…方面；以…措词
shield from 庇护使免遭

Notes

1. solar lean-to greenhouses：太阳能温室，就是利用太阳的能量，来提高塑料大棚内或玻璃房内的室内温度，以满足植物生长对温度的要求，所以人们往往把它称为人工暖房。

2. Zen：禅，是一种基于"静"的行为，源于人类本能，经过古代先民开发，形成各种系统的修行方法，并存在于各种教派。

3. The yin and the yang：阴阳是中国古代文明中对蕴藏在自然规律背后的、推动自然规律发展变化的根本因素的描述，是各种事物孕育、发展、成熟、衰退直至消亡的原动力，是奠定中华文明逻辑思维基础的核心要素。概括而言，按照易学思维理解，其所描述的是宇宙间的最基本要素及其作用，是伏羲易的基础概念。

Reading Comprehension

I. Each piece of the following information is given in one of the paragraphs in the passage. Identify the paragraph from which the information is derived and put the corresponding number in the space provided.

(   ) 1. Biological control of insects was used in the culture of orange and litchi (Zhou, 1998).

(   ) 2. These new greenhouses utilized advanced computerized control systems with automatic irrigation and environmental controls.

(   ) 3. The Chinese developed cultivation techniques like row planting to space out plant root zones.

(　　) 4. In the north, northeast, and northwest, where winter is cold, plastic tunnels are mainly used in early spring and late autumn.

(　　) 5. Later on during the Tang Dynasty (618—907), natural hot springs were used for forcing vegetables in the winter.

II. Decide whether the statements are true (T) or false (F) according to the passage.

(　　) 1. The ancient history of Chinese horticulture began along the fertile hills of the Yellow River around 2000 BC and the writings of Confucius (552 BC—479 BC) are the first source to mention horticulture.

(　　) 2. Tools such as wheelbarrows, pottery, and hoes were not utilized.

(　　) 3. Biological control of insects was used in the culture of orange and litchi.

(　　) 4. Simple greenhouses were made using translucent oiled paper as covers to grow vegetables and flowers during the Ming Dynasty.

(　　) 5. Modern greenhouses from foreign countries began to appear around China by the 20th century, they were made of glass, polycarbonate, or polyethylene, and had steel frames.

III. Discuss the following questions.

1. What else do you know about Chinese horticulture, especially the ancient Chinese horticulture?

2. Do you have more insights into Chinese horticulture, especially the ancient Chinese horticulture?

第四章　土壤耕作制度

Chapter 4　Soil Cultivation System

  山脊种植和围垄
Ridge Cultivation and Polder

Ridge cultivation, an ancient Chinese farming method, was used to grow crops on soil hilled above the ground. From the Qin Dynasty to the Jin Dynasty, as the increasing use of cattle farming and iron plow, China's soil tillage entered the age of ridge cultivation. This paralleled the till; widespread use of the mould board, which was particularly important, as well as the invention of rake and mo to work with other sorts of soil preparation tool.

A ridge is comprised of a high standing platform and a low lying furrow, looking similar to a wave. Thick soil with big voids, which are not easy to harden, is conducive to the growth of crop roots. Its cross-section approximates to an isosceles trapezoid. While in areas of Eastern China, is the ridge looks like a square. The platforms stand 16~20cm high and 60~70cm between each other. If the ridge distance is too far, crops cannot be closely planted; if too near, the ridge cannot resist drought and water logging. The surface area of the ridge increases by 20%~30% in tandem with the sunlit area, which consequently speeds up the heat absorption and dissipation. During the day, the ridge temperature is 2~3 degrees higher than that of the flat ground. The difference in height between the ridge and the furrow sets an ideal background for field drainage and drought prevention. Due to less soil moisture, the ridge structure becomes a favorable factor for potato tuber enlargement. The platform can block wind, reduce wind speed and wind erosion. The plant's higher base in the earth protects the crop from lodging and saves fertilizer during ridge tillage.

The methods of ridge building are as follows: (a) till the ridge after soil preparation because when the soil is broken up, it is easier for sowing or planting; (b) raise the ridge without prior soil preparation and leave the soil rough on the inside and finer on the outside with thin voids which are good for weathering; (c) build ridges of equal heights on the hillside and increase its depth to improve its fertility and water conservation capacity.

The ridge system originated in the Western Zhou Dynasty and became prevalent during the Warring States Period in the north where cold climate, spring drought and heavy summer rainfall were usual. In Sima Qian's Records of the Grand Historian—Basic Annals of Five Emperors, Emperor Shun appointed Hou Ji to take charge of agriculture. Hou Ji was born in the southwest of Wugong County, Shaanxi Province. He had developed a hobby of growing crops since a child. Thirteen years of experience in flood control with Da Yu deepened his understanding of the impact of drought and flood on agriculture. Legend tells that Hou Ji created the ridge technique. The details of ridge tillage were written in the book *Spring and Autumn Annals of Master Lv*, around 239 BC, in which three articles were considered as China's first papers on agriculture, namely *Land Use*, *On Earth* and *On Time*, each around 500 words. *Land Use* explained why it was important to build the ridge and how to use Leisi on the ridge. *On Earth* described ridge shapes and specifications. *On Time* emphasized the importance of farming time. The accurate documentation of the ridge system in *Spring and Autumn Annals of Master Lv* manifested the popularity of ridge farming in the Warring States Period.

Today in the north, northeast and Inner Mongolia, ridges are cultivated for maize, sorghum, sugar beet and other dryland crops. In other areas it is mainly used for sweet potatoes, taros and the like.

Polder

In the Song and Yuan Dynasties, agriculture developed to a new level. Population expansion, especially in the south, and land annexation put great pressure on agricultural production. Therefore, the expansion of arable land became an urgent priority. Compared with the north, the topography of the south was much more complex. In addition to early cultivated plains, there were more mountains and lakes. The solution to contend with this challenge was to use water and hills for farm land. It based on this societal and natural backdrop that numerous methods of field cultivation were invented.

In order to obtain more arable land, Chinese ancestors built up embankments along rivers, lakes and beaches, and surrounded water for land, finally developing one type of farmland called the diked field. It was also named as polder. This method began to be used in the late Spring and Autumn Period under the rule of the Wu and the Yue. The diked field had been established in Taihu Lake area, near Suzhou city where there was a large distribution of enclosed fields. In a diked field, an embankment was built to enclose the field. After Wu and Yue, in order to improve production and hydraulic technology, diked fields grew from being used sporadically to intensively. In the mid-Tang, citizens of the north migrated to the south in order to avoid wars, which resulted in a drastic leap in the southern population. The population growth increased the demand for food and arable land, which laid the foundation for the large-scale development of the diked field. Different from the initial one, it was developed into an integrated system combining the embankment, sluice and ditch. Each field stretched scores of li① and was built with internal canals and sluices. When drought came, the gates were opened to allow the river to flow for irrigation; when it flooded, the gates were closed to block the water. Farmers working in the diked field were no longer worried about drought or flood and instead, they gained great benefits from this system. Statistically, during 86 years of Wu and Yue history, the country was struck by only four floods and one drought with an average of once every 17 years. This was the lowest record in the Taihu Lake region, and yet at the same time harvests were very common. The downstream area of the Yangtze River witnessed rapid development after the Song Dynasty. As was written in Compendium Manuscript of Song Dynasty, in the Dangtu and Wuhu counties of Anhui Province, fields were diked with banks linked up to 480 li. In the first decade of Song Chunxi, there were 1489 diked fields around Taihu Lake, which played a considerable role in the expansion of cultivated land.

However, everything has its limit. In the Song Dynasty, to satisfy the insatiability of some bureaucrats and landlords, the rivers and fields were so blindly exploited that lake surface area shrank, waterways became obstructed, irrigation failed and the water environment was extremely damaged. This exacerbated flooding and droughts with agricultural land in the area decreasing dramatically. The Song government was forced to constantly provide public relief, plunging it into grave financial circumstances. Additionally, in the Southern Song Dynasty, Lian Lake was almost drought up, storing little water in either spring or summer, it had lost its capacity of regulating canal waters, which certainly had an impact on southern canal transport. This is an extremely profound and negative historical lesson. The key to bring a positive result lies in appropriate exploitation and sound governance.

Words & Expressions

ridge/rɪdʒ/*n.* 山脊,山脉
polder/ˈpəʊldə(r)/*n.* 开拓地;围垦地
bureaucrat/ˈbjʊərəkræt/*n.* 官僚,官僚主义者

Notes

polder:在沿江、滨湖或沿海滩地上建造围堤阻隔外水进行垦殖的工作。滨湖围垦叫作"围湖造田",海滩地围垦叫作"海涂围垦"或"围海造田"。围垦区内应建有完整的排灌系统,以保证区内土地的正常开发利用。

---

① li,里,中国古代使用的长度计量单位,1 里=500 米。

Reading Comprehension

I. Each piece of the following information is given in one of the paragraphs in the passage. Identify the paragraph from which the information is derived and put the corresponding number in the space provided.

(   ) 1. Thick soil with big voids, which not easy to harden, is conducive to the growth of crop roots.

(   ) 2. During the day, the ridge temperature is 2~3 degrees higher than that of the flat ground.

(   ) 3. The ridge system originated in the Western Zhou Dynasty and became prevalent during the Warring States Period in the north where cold climate, spring drought and heavy summer rainfall were usual.

(   ) 4. Today in the north, northeast and Inner Mongolia, ridges are cultivated for maize, sorghum, sugar beet and other dry land crops.

(   ) 5. In the Song and Yuan Dynasties, agriculture developed to a new level. Population expansion, especially in the south, and land annexation put great pressure on agricultural production.

II. Decide whether the statements are true (T) or false (F) according to the passage.

(   ) 1. This method began to be used in the late Qin Period under the rule of the Wu and the Yue.

(   ) 2. The plant's higher base in the earth protects the crop from lodging and saves fertilizer during ridge tillage.

(   ) 3. In the first decade of Song Chunxi, there were 1889 diked fields around Taihu Lake, which played a considerable role in the expansion of cultivated land.

(   ) 4. Statistically, during 86 years of Wu and Yue history, the country was struck by only four floods and one drought with an average of once every 17 years.

(   ) 5. Due to less soil moisture, the ridge structure becomes a favorable factor for potato tuber enlargement.

Language Focus

III. Complete the sentences with the correct form of the words in the table.

absorption	circumstance	dramatically	solution	difference
technology	Cultivation	downstream	ancestor	expansion

1. Ridge_____, an ancient Chinese farming method, was used to growing crops on soil hilled above the ground.

2. The surface area of the ridge increases by 20%~30% in tandem with the sunlit area,

which consequently speeds up the heat_____ and dissipation.

3. Population_____, especially in the south, and land annexation put great pressure on agricultural production.

4. This exacerbated flooding and droughts with agricultural land in the area decreasing _____.

5. After Wu and Yue, in order to improve production and hydraulic_____, diked fields grew from being used sporadically to intensively.

6. In order to obtain more arable land, Chinese_____ built up embankments along rivers, lakes and beaches, and surrounded water for land, finally developing one type of farmland called the diked field.

7. The_____ to contend with this challenge was to use water and hills for farm land.

8. The Song government was forced to constantly provide public relief, plunging it into grave financial.

9. The_____ area of the Yangtze River witnessed rapid development after the Song Dynasty.

10. The _____ in height between the ridge and the furrow sets an ideal background for field drainage and drought prevention.

IV. Match the sentences in Section A with the English translation in Section B.

Section A
1. 除了早期开垦的平原，还有更多的山脉和湖泊。
2. 植物较高的基地在土壤中保护作物不倒伏和节省肥料在垄耕作。
3. 山脊和犁沟之间的高度差异为农田排水和防旱提供了理想的背景。
4. 这加剧了洪水和干旱，该地区的农业用地急剧减少。
5. 山脊由高高的平台和低矮的犁沟组成，形似海浪。

Section B
1. A ridge is comprised of a high standing platform and a low lying furrow, looking similar to a wave.
2. The plant's higher base in the earth protects the crop from lodging and saves fertilizer during ridge tillage.
3. In addition to early cultivated plains, there were more mountains and lakes.
4. The difference in height between the ridge and the furrow sets an ideal background for field drainage and drought prevention.
5. This exacerbated flooding and droughts with agricultural land in the area decreasing dramatically.

V. Translate the paragraph into English.

田垄耕作是中国古代的一种耕作方法，用来在地面上碾压土壤上种植作物。从秦朝到晋代，随着耕牛和铁犁的使用越来越多，中国的土壤耕作进入了垄耕时代。与耕作同时利用，特别重要的是组合犁壁，以及利用耙子和钼与其他类型的土壤准备工具一起协作。

Development

VI. Discuss the following questions.
1. What's the limitation of polder in the Song Dynasty?
2. How did agriculture develop to a new level in the Song and Yuan Dynasties?

复种梯田
Multiple Cropping and Terraces

Multiple Cropping

Multiple cropping is a cropping system that grows crops on the same piece of land more than once within a year. For example, growing crops two or three times a year is called double cropping or the one-year-three-cropping system; three times in two years is called the two-year-three-cropping system, which are generally termed as multiple Cropping. In China, multiple cropping began in the Spring and Autumn Period 2000 years ago and it was an important part of the intensive farming technology. Crop rotation is planting different crops at different times on a piece of land in a certain order. From experience, multiple cropping proved highly efficient in both land and sunlight utilization, and significantly beneficial in per unit yield increase.

In the Qin and Han Dynasties, rotation and multiple cropping began to take shape. In Rites of Zhou, a classic in Eastern Han Dynasty, Zheng Xuan annotated for "field keeper" and "weeder". He said that in Eastern Han, millet planting was succeeded by wheat planting after the harvest, or after the first wheat harvest it was suitable to plant millet and beans. In other words, crop rotation and multiple cropping between cereal, wheat and bean had already been well implemented in the Qin and Han Dynasties along the middle and lower reaches of the Yellow River. In the Wei and Jin Dynasties, crop rotation had also seen its initial development in south China. Wudu Prose by Zuo Si and Guang Zhi by Guo Yigong both documented such a planting system.

The system was comprehensively discussed in the *Arts for the People* by the famous agronomist Jia Sixie. He emphasized that most crops needed a reasonable cropping rotation. In this book, he set the rules for alternatively growing bean and cereal. He even summarized the green manure cropping experience in the north.

In the Ming and Qing Dynasties, China's cropping rotation developed into an integrated new stage where rotation cropping, intercropping and interplanting came together. In the Tang Dynasty, the Chinese economic center moved southward where double cropping of rice cultivation had greater development. In the Song Dynasty, the output of wheat was promoted in the area where rice, wheat and other plants were alternately grown with two harvests a year, and the total output was constantly expanding. In the Southern Song Dynasty, multiple cropping of rice, bean, wheat and vegetables developed in the south. Three harvests a year of rice by interplanting were written in Knowledge of the Southern Five Ridges by Zhou Qufei. In the Ming Dynasty, the outstanding agronomist Xu Guangqi summarized the following practices in *Agricultural Policy Book*:

It was best to plant cotton once a year, and plant cotton and wheat by double cropping when it had to, for the case of which only barley, highland barley would work, for the two were harvested early just in time for sowing cotton seeds.

In the fields where both cotton and rice were grown, it would be better to grow two years of cotton, then one year of rice so that the soil would become rich again and few insects bred after the grass roots rot.

Rotation cropping and multiple cropping technologies are the essence of Chinese traditional agriculture. They are worthy of thorough studies. China's population is gradually on the rise whereas the size of arable land moves in the opposite direction. In this case, the multiple cropping is still an indispensable measure in China to improve the utilization of agricultural land and especially to increase agricultural production per unit area.

Sandy Field

The Longzhong region has a typical continental climate with an annual rainfall of only 300mm, while evaporation is as high as 1500mm. It is rather dry and extremely unfavorable for crop growth. Since ancient times, the natives of this region paid much attention to water conservation. Several counties in Gansu Province of central China have fields covered with sand and pebbles so as not to leave any soil uncovered while seedlings grow. The sandy field was originated from the mid-Ming Dynasty, and is still used today. Its construction could be divided into two steps: (a) the sand was dug up and then placed aside. The Longzhong area was full of alluvial deposits where pebbles and grit were buried. Sand could be dug up everywhere, which became an inexhaustible source for sandy field construction. The laying of sand usually started in September and continued to next March. To start, the land should be thoroughly plowed, fertilized, leveled and compacted. Generally, the thickness of the sand layer stretched four to five inches. The sand was then spread evenly. If sand accidentally mixed with dirt, the sandy field would lose its properties. It could be used for 40 years at most, gradually after which, the sandy field would deteriorate seriously and the output would be less than the average. (b) To prevent this, old sand should be cleared and replaced with new sand. There is a popular old saying in Longzhong: "too much labor for grandpa, too much wealth for son, too much hunger for grandson. " It indicates that accumulating and laying of sand were of heavy workload, which is a drawback of sandy land.

The sandy field is able to increase production for the following reason:

Water conservation: when it rains, large voids form between the grit and allow water to filter into the soil. The gravel holds back the gravel rainwater and reduces surface runoff. Meanwhile, separated by gravel, the top soil is not directly exposed to the sun, because the soil moisture is contained.

Soil conservation: Longzhong is located in the Loess Plateau where soil erosion occurs frequently. When the field surface is covered by gravel, soil erosion caused by rain and strong winds can be effectively prevented. In this sense, the sandy field is of great significance in Longzhong.

The insulation: the temperature in Longzhong is relatively low and the daily range of it is up to 20 degrees. The top grit mitigates temperature differences, which is quite helpful for the growth of crop. In sandy field, seeds can germinate earlier and crops ripen faster. Thus the harm from early frost is effectively lessened. For example, due to low temperatures and less rainfall, Lanzhou was not originally a suitable place for cotton plantation at first. But this has become a story of the past due to the insulation effect of the sandy field.

Lower alkali: the soils in Longzhong are mostly alkaline. Because of this, sandy land functions well in water conservation. Excessive water can dissolve during permeation while diluting the saline and alkaline soil in the plowed layer. Therefore, the soil salinity and alkalinity are lowered.

Easy management: because the field is covered by sand, few weeds survive, requiring weeding only once.

These farming techniques have been playing an important role for 400~500 years. So far, the sandy field is still the main farming method for local people to combat drought.

Words & Expressions

rotation/rəʊ'teɪʃn/n. 轮作，轮耕
intercropping/ɪntə'krɒpɪŋ/n. ［农学］间作；［农学］间混作
agronomist/ə'grɒnəmɪst/n. 农业师；农学家；农艺师
conservation/ˌkɒnsə'veɪʃn/n. 保护，保存；节约，防止浪费；守恒定律
alkali/'ælkəlaɪ/n. 碱；可溶性无机盐

Notes

1. multiple cropping：复种是在同一耕地上一年种收一茬以上作物的种植方式。有两年播种三茬，一年播种二茬，一年播种三茬等复种方式。复种的方法有：复播，即在前作物收获后播种后作物；复栽，即在前作物收获后移栽后作物；套种，即在前作物成

熟收获前，在其行间和带间播入或栽入后茬作物。扩大复种面积，提高复种指数，是充分利用气候资源和土地资源，提高单位耕地面积产量和总量的有效措施。

2. sandy field：（土地科学术语）沙地，指表层被沙覆盖、基本无植被的土地，包括沙漠，不包括水系中的沙滩。

Reading Comprehension

I. Each piece of the following information is given in one of the paragraphs in the passage. Identify the paragraph from which the information is derived and put the corresponding number in the space provided.

(　　) 1. For example, growing crops two or three times a year is called double cropping or the one-year-three-cropping system.

(　　) 2. In the Qin and Han Dynasties, rotation and multiple cropping began to take shape.

(　　) 3. The sandy field originated from the mid-Ming Dynasty, and is still used today.

(　　) 4. Excessive water can dissolve during permeation while diluting the saline and alkaline soil in the plowed layer.

(　　) 5. In the Ming and Qing Dynasties, China's cropping rotation developed into an integrated new stage where rotation cropping, intercropping and interplanting came together.

II. Decide whether the statements are true (T) or false (F) according to the passage.

(　　) 1. Generally, the thickness of the sand layer stretched one to five inches.

(　　) 2. When the field surface is covered by gravel, soil erosion caused by rain and strong winds can be effectively prevented.

(　　) 3. China's population is gradually on the rise whereas the size of arable land moves in the opposite direction.

(　　) 4. Easy management: because the field is covered by sand, many weeds survive, requiring weeding only once.

(　　) 5. In the fields where both cotton and rice were grown, it would be better to grow two years of cotton, then one year of rice so that the soil would become rich again and few insects bred after the grass roots rot.

Development

III. Discuss the following questions.

1. What is the multiple cropping?
2. Why is the sandy field able to increase production

精耕细作体系
Intensive Cultivation System

Terraces

After the mid-Tang Dynasty, a large percentage of the population migrated to the south, leading to the explosion of population in the southern, under the condition of which, people started farming on hills. During Kaiyuan and Tianbao years of the Tang, people built sloped fields on the hills and called it She field (畲田). To do this the woods and weeds were cut down by broad knives and burnt for fertilizer. Then the crops were planted along the slopes.

However, heavy rainfall washed away lots of soil, causing severe soil erosion. Thus the hills were no longer suitable for cultivation in the third year. In order to ease the conflict between hill tillage and soil erosion, farmers in the Song Dynasty improved the terrace technology. In fact, terraces could be traced back to an earlier era. A clay model unearthed in an Eastern Han grave in Pengshui County, Sichuan provides such evidence. In this model, the space between two hillocks were narrow and striped, shaped like fish scales. The hill structures shared a slight resemblance to staircases. The clay model was quite similar to today's Sichuan terraces. It can be inferred that this was a primitive state in terrace development with its prototype created in the Han Dynasty. It was not exactly farming along the hillside, but on horizontal steps. Additionally, each step was fortified by a low stone or earthen bank, which prevented water and soil loss. According to Wang Zhen, there were three types of terraces: (a) if the mountain contained only soil, the farmer would build up solid steps from the bottom to the top; (b) if the mountain was of half soil and half stone, then stone bases would be required to surround the soil; (c) if it was a steep mountain terrace with a water source, then it was recommended to grow rice. If without, then grow dry land crops such as millet or wheat.

Terraces rely mainly on natural rainfall for irrigation. In order to use the limited water resources, people after the Song Dynasty took certain measures to construct ponds on higher ground where the water source was relatively intensive. Using the proportion of two to three mu[①] per 10 ponds were excavated for water storage. The edges of the ponds were required to be tall and the inside to be deep which enables the ponds to reserve more irrigation water against drought and to prevent crops from floods by impounding more flood water. Another benefit was that mulberry and cudrania trees could be planted on the tall embankments to which cattle were tied and under which they were shaded. The embankment soil was compacted when the cattle trampling and the trees were fertilized by the dung of the cattle.

---

① mu，亩，中国古代面积单位，1 亩≈0.0667 公顷。

In the Song and Yuan Dynasties, the high speed waterwheel further promoted the development of terraces. This machine drew water up to the mountain to irrigate terraces. For high mountains, two waterwheels were relayed, supplemented with a pond in between. Compared with the She field which prevailed in the Tang Dynasty, the terraces went much further in improving the effect of soil, water and fertilizer conservation. The topography of mountain took up two-thirds of the total southern area, therefore the adoption and spread of terraces was of great significance.

Though farming on hills had its advantages, it must be scientifically controlled. Once over exploited, it would become counterproductive. In the Qing Dynasty, it was recorded in *The Book of Shed Dwellers* some of those devastating consequences. Today, terraces are still in use in some mountain areas.

Shelf Field

Most people in China may think that artificial floating land, like gardens sitting on the water, are foreign concepts in modern times. Actually, such creations were used by the farmers in ancient China. This production method was called the shelf field. The shelf field was also named the feng field. Historically, wild rice roots were called feng. A variety of water plants such as wild rice and cattail grew along southern rivers and lakes, flourishing and withering every year with their vines tangling together, hence the name feng field was adopted. Soaked in water and then blown by wind, the roots of the water weeds were gradually separated from the mud and formed into a small floating island several feet thick and dozens of feet wide. Farmers shoveled off the stems and leaves and then planted rice and vegetables. However, it took many years to form a natural feng field. To deal with this problem, farmers built wooden frames that filled with muddy grass roots as artificial feng field. To prevent it from drifting away or being stolen, it was tied by a rope to the shore. When the weather was bad, farmers pulled it away by hands or boat to the shelter until the weather cleared. In the Yuan Dynasty the feng field was officially named the shelf field.

The shelf field developed drastically in the Song and Yuan Dynasties. Compared with the diked field, the shelf field had quite a few merits: (a) it did not take up too much space and did not decrease the surface area of the water; (b) it could vary according to production need and water area; (c) it fluctuated with the water thus wouldn't be threatened by flood and drought; (d) it did not do harm to fishing, to the contrary, it contributed to the development and utilization of natural resources and safeguarded the eco-balance; (e) it required little investment but provided high economic reward.

Zone Field Method

The zone field method, also called zone cropping, a cultivation method that holds soil moisture and resists drought. It was invented by the well-known agronomist Fan Shengzhi during Emperor Cheng's reign of the Han Dynasty. The field zones were arranged in two forms: (a) broad spaced sowing, e.g. to sow seeds in the shallow furrow on fat ground; (b) to square spaced sowing, e. g. dug small square zones, the size and depth of which depended on the fertility of the soil and the crop variety. In general, on fertile land, the zone should be about 12 inches in length, width and depth, and every two zones should be about 12 inches in between, while on barren land, the zone interval extended further. In other words, the more fertile the soil was, the more zones fields there were, and vice versa.

Zone fieldmethod had many different features. To start with, they were unrestricted by soil, meaning they could be utilized on infertile land, which not only expanded land application but also did good to soil and water conservation. Another feature was that instead of plowing every inch of land, farmers dug deeply, and then plowed, applied fertilize, watered and weeded within the zone. Intertillage was practiced in confined areas so that fertilizing and watering could be more economical and focused. Intensive farming in small areas helped plants become more drought- resistant and made high yield possible in a small area.

The zone field method had some limitations as well. It was labor intense, time consuming and unsuitable for the use of livestock due to limited traction. As a result, it was not promoted at the national level.

Field Substitution Method

The field substitution method was a pioneering farming method adapted in the northern uplands and promoted by Zhao Guo (the officer in charge of agriculture during the Western Han Dynasty). Since the planted ridges on the same piece of land were changed every other year, this method was named the field substitution method.

According to *The Book of Han—Agriculture Economy and Currency Annals* (《汉书·食货志》), the field substitution method could be generalized as follows:

(1) In one mu reclaim three furrows and three ridges, with each one of them one foot wide.

(2) In the first year, plant crops in the ditch, allow weed to burgeon in the ridges, and constantly supplement the roots with soil from the ridges until summer when the ridges are leveled.

(3) In the second year, ditch the previous ridges, ridge the previous ditches, and plant the same way.

What are the advantages of field substitution? Firstly, land in production and land in recovery progress alternately. In case of fertilizer shortage, soil fertility could acquire natural restoration and enhancement. Secondly, in the ditches, due to less soil evaporation and ridges as wind barriers, the seeds are laid in a favorable seedling environment. Thirdly, weeding plus cultivation lead the crop roots to extended growth, and therefore the plant could absorb more fertilizer and more moisture. It also better withstands wind, drought and lodging, "Less force but more produce" that's how Ban Gu summarized the practical effect of field substitution. Combined with coupled plow, seed plow and other tools, labor productivity was remarkably improved. The field substitution method was spread from the Central Shaanxi Plain to broader regions including Juyan (now Inner Mongolia Ejinaqi area), as well as in Henan, Shanxi, and northwest Gansu. The implementation of this method achieved remarkable results—the increasing of crop yield and cultivated field areas. It also played an important role in social and economic recovery in the later ruling period of Emperor Wu of the Han Dynasty.

However, in the late Western Han Dynasty, this method was no longer visible in *The Book of Fan Shengzhi*. For one thing, it was very demanding on cattle and farm tools. Thus, it was more suitable for tillage on a large scale, whereas under China's feudal agricultural background, the small scale decentralized operations lacked adequate capacity to optimally utilize the system. It was likely that only boarder farmland, government public land and large farms of the rich and powerful families were capable of applying the field substitution method. Though not practicing for long, its advanced technology was inherited by later generations and had a profound impact on the development of China's agricultural science and technology.

Words & Expressions

terraces/'terəsɪz/*n*. ［地质］阶地；［农］梯田
furrow/'fʌrəʊ/*n*. 皱纹；犁沟；车辙 *vt*. 犁；耕；弄绉 *vi*. 犁田；开沟；犁出浪迹
vice versa *adv*. 反之亦然
substitution/ˌsʌbstɪ'tjuːʃn/*n*. 代替，替换；代替物

Notes

1. She field：畲田，是指采用刀耕火种的方法耕种的田地。
2. Shelf field：葑田有两个意思：一是湖泽中葑菱积聚处，年久腐化变为泥土，水涸成田，是谓"葑田"；二是将湖泽中葑泥移附木架上，浮于水面，成为可以移动的农田，叫葑田，也叫架田。
3. *The Book of Han—Agriculture Economy and Currency Annals*：《汉书·食货志》，分上下两卷，上卷谈"食"，即农业经济状况；下卷论"货"，即商业和货币情况，是当时的经济专篇。

Reading Comprehension

I. Each piece of the following information is given in one of the paragraphs in the passage. Identify the paragraph from which the information is derived and put the corresponding number in the space provided.

(   ) 1. To start with, they were unrestricted by soil, meaning they could be utilized on infertile land, which not only expanded land application but also did good to soil and water conservation.

(   ) 2. The shelf field developed drastically in the Song and Yuan dynasties.

(   ) 3. In the Qing Dynasty, it was recorded in *The Book of Shed Dwellers* some of those devastating consequences.

(   ) 4. Since the planted ridges on the same piece of land were changed every other year, this method was named the field substitution method.

(   ) 5. However, heavy rainfall washed away lots of soil, causing severe soil erosion.

II. Decide whether the statements are true (T) or false (F) according to the passage.

(   ) 1. However, in the early Western Han Dynasty, this method was no longer visible in *The Book of Fan Shengzhi*. For one thing, it was very demanding on cattle and farm tools.

(   ) 2. It was labor intense, time consuming and unsuitable for the use of livestock due to limited traction.

(   ) 3. Most people in China may think that artificial floating land, like gardens sitting on the water, are foreign concepts in modern times.

(   ) 4. In order to ease the conflict between hill tillage and soil erosion, farmers in the Qin Dynasty improved the terrace technology.

(   ) 5. A variety of water plants such as wild rice and cattail grew along southern rivers and lakes, flourishing and withering every year with their vines tangling together, hence the name feng field was adopted.

III. Discuss the following questions.

1. How did people start farming on hills?

2. According to *The Book of Han—Agriculture Economy and Currency Annals*, how did the field substitution method be generalized?

第五章　传统农业灌溉技术

Chapter 5　Irrigation

Text A　古代农业灌溉　Ancient Irrigation

Foot-powered irrigation Agriculture and irrigation account for large amounts of water use. Water for irrigation can get from wells, rivers, canals, lakes, ponds and reservoirs. Often dams are built to supply water for irrigation. There is 545960km^2 of irrigated land in China. 40% of China's crop land is irrigated, compared to 23% in India. The average yield per acre in China is double that of India. Chinese invented the potters wheel, escapements (important parts of a watch that keep the springs from unwinding too fast when they are tightly wound), the first computer, the first canal lock, deep drilling devices, efficient animals harnesses, and the first true mechanical crank.

Pumps are important for irrigation. In the old days water wheels and manual labor were needed to lift water from wells, rivers, canals and ponds to agricultural land. Now gasoline and diesel-powered pumps do much of the work. Pumps may be noisy but are a relatively cheap and efficient.

Studies in South Asia have shown that a traditional treadle water pump—operated by a person on a device that looks a bit like a stair-master , stair climber—can increases the income of farmers by 25%. First introduced to Bangladesh in the 1980s and now widely used in Asia and sub-Sahara Africa, these pumps are easy to install and simple to operate and often deliver higher crops yields than those using diesel pumps.

Water used in irrigation often originates from sources miles away and sometimes uses tunnels, aqueducts and canals that were built over 1000 years ago. The water often flows from the source in a single canal which in turn divides in smaller canals that lead to the fields.

The water flowing in and out is regulated by a complex process that has been fine tuned over the centuries. In many places water is distributed by opening and closing gates which provide water to certain area for a specific amount of time. Some places still use water clocks (a pot with a small hole in the bottom that measures out about three minutes) to determine when the gates of the irrigation system open.

The schedule for distributing water is May by an annual calendar. The various stages of growing process—the flooding of the fields, the transplanting of sampling and the harvesting of the nature crop—are all determined by the calendar and sometimes each stage is marked by festivals and rites.

The distribution of water is often overseen by individuals who follow customs and rules to determine which fields get water. In traditional China the water is often distributed through a system based on fairness and equity in turn based on family size and land holdings. Each member of the community is entitled to their share as long as they fulfill their duties in maintaining the system. In corrupt system, water is often diverted to large landowners who produce cash crops which the government receives taxes or kickbacks.

Irrigation with canals is very inefficient. Lots of water is lost to evaporation, runs off and absorbs into the soil before it reaches crops. Governments are often to blame for these practices because they subsidize water so heavily that farmers have little incentive to save it.

Poorly-drained irrigated land leaves behind salt deposits as water evaporates. In many places, fields that once grew bountiful crops of grains are now encrusted in salt. More than a quarter of the world's irrigated land has become so salty that many crops will no longer grow there. To make the land productive again the fields have to be flooded four times to clear away the salt.

Irrigation also causes large amount of salts, fertilizers and pesticides to be flushed into rivers and streams.

The drilling of wells for irrigation, farming and animal herding can trigger an unhealthy cycle. Drilling wells causes the water table to drop. After a while the water may become too salty for crops and animals or too expensive to pump resulting in the sinking of more wells, which causes the water table to drop further.

Words & Expressions

irrigation/ˌɪrɪˈgeɪʃn/ *n*. 灌溉
reservoir/ˈrezəvwɑː(r)/ *n*. 水库，蓄水池
acre/ˈeɪkə(r)/ *n*. 土地，地产；英亩
harness/ˈhɑːnɪs/ *n*. （马的）挽具，马具
diesel/ˈdiːz(ə)l/ *n*. 柴油；内燃机，柴油车
treadle/ˈtredl/ *n*. 踏板；踏木
aqueduct/ˈækwɪdʌkt/ *n*. [水利] 渡槽；导水管；沟渠

evaporation/ˌɪvæpə'reɪʃn/ n. 蒸发

Notes

Water Clock：水钟，在中国又叫刻漏、漏壶。根据等时性原理滴水记时有两种方法，一种是利用特殊容器记录把水漏完的时间（泄水型）；另一种是底部不开口的容器，记录它用多少时间把水装满（受水型）。中国的水钟，最先是泄水型，后来泄水型与受水型同时并用或两者合一。自公元 85 年左右，浮子上装有漏箭的受水型漏壶逐渐流行，甚至到处使用。

Reading Comprehension

I. Each piece of the following information is given in one of the paragraphs in the passage. Identify the paragraph from which the information is derived and put the corresponding number in the space provided.

(　　) 1. Water used in irrigation often times originates from sources miles away and sometimes uses tunnels, aqueducts and canals that were built over 1000 years ago.

(　　) 2. Foot-powered irrigation agriculture and irrigation account for large amounts of all water use.

(　　) 3. In traditional societies the water is often distributed through a system based on fairness and equity in turn based on family size and land holdings.

(　　) 4. In the old days water wheels and manual labor were needed to lift water from wells, rivers, canals and ponds to agricultural land.

(　　) 5. After a while the water may become too salty for crops and animals or too expensive to pump resulting in the sinking of more wells, which causes the water table to drop further.

II. Decide whether the statements are true (T) or false (F) according to the passage.

(　　) 1. Foot-powered pumps, using 1000 years old design are being resurrected and used not only in China but in Africa and other parts of world as an ecofriendly that doesn't require a lot of energy and draws water in a sustainable way so as not threaten groundwater supplies.

(　　) 2. In many places water is distributed by opening and closing gates which provide water to certain area for a specific amount of time.

(　　) 3. The schedule for distributing water is may by an annual calendar.

(　　) 4. In the old days water wheels and manual labor were not needed to lift water from wells, rivers, canals and ponds to agricultural land.

(　　) 5. Irrigation with canals is very efficient.

Language Focus

III. Complete the sentences with the correct form of the words in the table.

| distribution | fertilizer | average | expensive | productive |
| corrupt | square | subsidize | irrigate | schedule |

1. There is 545960 _____ kilometers of irrigated land in China.

2. There may thousands of fields, each _____ about from a few acres to may dozen acres in size, or huge swaths of agricultural land.

3. More than a quarter of the world's _____ land has become so salty that many crops will no longer grow there.

4. The _____ of water is often overseen by individuals who follow customs and rules to determines which fields get water when.

5. Irrigation also causes large amount of salts, _____ and pesticides to be flushed into rivers and streams.

6. After a while the water may become too salty for crops and animals or too _____ to pump resulting in the sinking of more wells, which causes the water table to drop further.

7. Governments are often to blame for these practices because they _____ water so heavily that farmers have little incentive to save it.

8. To make the land _____ again the fields have to be flooded four times to clear away the salt.

9. In _____ system, water is often diverted to large landowners who produce cash crops which the government receives taxes or kickbacks.

10. The _____ for distributing water is May by an annual calendar.

IV. Translate the paragraph into English.

Section A

1. 中国的灌溉耕地占总数的40%，相比较印度的这一比例是23%。
2. 社会上的每个人都有权利分享自己的一份，只要他们履行了维护这个制度的责任。
3. 水的流入和流出是由一个复杂的过程调节，几个世纪以来一直在运转良好。
4. 按时间表分配水的时间是五月。
5. 政府经常因为这些做法而受到指责，因为他们补贴水太多，以至于农民没有动力去节约用水。

Section B

1. The water flowing in and out is regulated by a complex process that has been fine tuned over the centuries.
2. Governments are often to blame for these practices because they subsidize water so heavily that farmers have little incentive to save it.
3. 40% of China's crop land is irrigated, compared to 23% in India.
4. The schedule for distributing water is May by an annual calendar.

5. Each member of the community is entitled to their share as long as they fulfill their duties in maintaining the system.

V. Translate the paragraph into Chinese.

The schedule for distributing water is May by an annual calendar. The various stages of growing process—the flooding of the fields, the transplanting of sampling and the harvesting of the nature crop—are all determined by the calendar and sometimes each stage is marked by festivals and rites.

_____
_____
_____
_____
_____

VI. Discuss the following questions.

1. After reading the text do you have more insights in the irrigation in ancient agriculture in China?

2. How did the water distribute in ancient China?

 旱地水资源保护
Water Conservation on Dry Land

China's Yellow River basin is a semi-arid, sub-humid region where the annual rainfall reaches only 400~750mm. Additionally, the distribution of rainfall is very uneven, of which 60% occurs in summer and autumn. It is the main planting season in spring. However, rainfall there remains very scarce, coupled with windy weather, often resulting in a poor harvest. In the long term struggling with drought, various soil moisture conserving and water-saving technologies were created by northern farmers. Dry land farming techniques of the Yellow River Basin was a typical example. In this region, people had to cover seeds immediately after sowing, or the seeds might quickly be dried in the sun or blown away by the wind. The technology of deep plowing and fine harrowing was created to solve this problem. Deep plowing not only improved soil absorption capacity, but also helped plants grow more firmly and take in more water and nutrients. What's more, after the deep plowing, the pests hidden underground were largely exposed which allowed their predators to hunt them more easily. During the same process, the weeds, another enemy to the crops, were not to safe, for their roots were generally damaged by the mechanical labor. Fine harrowing occurred soon after plowing to refine the soil so as to reduce evaporation and maintain soil moisture. The methods of deep plowing and fine harrowing were great development in water conservation and drought prevention in ancient China.

In the Han Dynasty, moisture conservation tillage technology progressed further as the importance on knowing well the right timing was attached. It was pointed out in *The Book of Fan Shengzhi* that there were three ideal farming timings: the early spring thaw when the soil was loose and soft; summer solstice when the weather turned hot with increased rainfall and the soil was getting prepared; autumn equinox when both the climate and the soil were fine. Conversely, farming at the wrong time was also documented in the book, which led to predictable disasters. Before or missing the spring thaw, when the arable land breathed freely, plowing would cause moisture loss. In autumn, if farmers plowed without rainfall, soil moisture would be depleted, resulting in stiff soil layers. This was called withered autumn field. Plowing on wintry days also incurred ground water loss. The soil then turned dry and tough, hence the barren winter field. Additionally, the moisture conservation under snow became an important technique for agriculture. People of the Han Dynasty knew how to make use of snowfall to keep soil moist, due to the fact that on the one hand, the accumulated snow formed a perfect insulation on the top ground, which prevented the soil from being frozen so that plants could survive the coldness; on the other hand, very little water evaporated from the snow-capped ground. The above examples mentioned reflect how farmers of the Han Dynasty had accumulated a wealth of ground water protection experience and knowledge.

Since the Han and Wei Dynasties, due to the application of cattle farming, iron plowing and the emergence of the animal hauled leveler, the ta tillage (踏耕) method could be repeated before sowing. After each plowing, a rake was used to break the earth, followed by the mo which further refined and flattened the topsoil to prevent water evaporation. As a result, the soil structure was loose in the top layer and solid at the bottom, which preserved water and manure fertilizer well. Meanwhile, autumn farming was a good way to save water for spring in the next year. "Plow—harrow—level—press—hoe" was the main procedure for the dry land farming system practiced in the Yellow River Basin. With this set of sophisticated and ingenious farming technologies, the threat of spring drought was largely mitigated in this region.

Between the Han Dynasty and the North Wei Dynasty, in addition to water conservation techniques, seed selection technology made significant progress; numerous crop varieties were bred; pest control techniques and natural disaster management experienced considerable achievements. The classic work *Arts for the People* was written during this time, which represented the highest level of world agronomy and marked the formation of China's intensive agricultural technology system.

Scooping bucket

5000~6000 years ago, Chinese ancestors learned simple techniques to deal with field irrigation, namely a man scooping a jug of water from the river to irrigate the fields one after another. Ancient books recorded a method called "dig tunnel to the well, carry urn to irrigate".

After a long period of practice, the scooping bucket technique for irrigation was created. The bucket was a keg with ropes tied to both sides. In the north, most of the keg was made of wood, while in the south it was made of wicker. It required two men to pull the rope, then to throw the water ashore.

Shadoof

In the Spring and Autumn Period, the shadoof (also named "balance arm") technique was developed based on the practice of leverage theory. It was an erected shelf or a tree appended with, a slender lever, the middle of which was a fulcrum. The end of it was suspended with a heavy weight and a barrel hung in the front. After the barrel was lowered to fill with water, it could be easily hoisted to the desired height due to the weight of the lever end. It was written in *Garden of Stories*—Fan Zhi that in the Spring and Autumn Period, Deng Xi taught Wei citizens how to make use of balance arm, which increased water use efficiency from one section of the field to one hundred per day.

Windlass

As early as 1100 years ago, the windlass was invented. It was a water-fetching apparatus evolved from a lever. A wooden frame was built above a well with a long tube outside along the horizontal axis. A long rope was wrapped around the tube with one end of the rope for hanging the bucket. At the head of the tube a crank was fixed, and it controlled the up and down motion of the rope and bucket. In this way, the water was lifted upward. Basically, the windlass shared a similar structure with pulley, which saved a lot of effort. In the Spring and Autumn Period, it had become a very popular technique in agricultural irrigation. It was written in volume Ⅲ of *Arts for the People* that three or four windlasses above deep wells could irrigate a large area of plowland. Now, in some mountain areas where groundwater extends deep down, the windlass is still used for obtaining drinking.

Chain pump

Thanks to the rolling wheel, it was possible to invent chain pump. It made an enormous contribution to China's agricultural development by satisfying the need for large scale field irrigation.

The earliest mention of the rolling wheel was seen in *The Book of the Later Han—Biography of Zhang Rang* during the rule of Emperor Xian of the Han Dynasty. Bi Lan, an eunuch official, might be the father of the rolling wheel. Then during the Three Kingdoms Period, Ma Jun from the Kingdom of Wei made further improvements, which not only raised its irrigation efficiency, but also simplified its structure so that it became fully operational even for children. The improved rolling wheel could be driven by hand, foot, cattle, water or wind. The transverse planks functioned as a chain, lying in a long rectangle groove with the body tilted by the river or the pond. The lower chain wheel and part of the body were submerged in the water. When the chain wheel was pushed, the plank impellers carried water up from the groove to the upper part of the wheel and then poured it out as the wheel came down.

In the Song Dynasty, the chain pump was widely employed in drought control, highland irrigation and low field drainage, mostly in Zhejiang Province, south of the lower reaches of the Yangtze River, Huainan, Fujian Province and other places. With marvelous pumping capacity, it operated and transported water in a very convenient and efficient way. Historically, the chain pump played such an important role that it was defined as one of the most extraordinary irrigation machines in China.

Waterwheel

The waterwheel is another type of irrigation machine created in the Sui and Tang Dynasties. It is completely hydraulic powered by the turning wheels. In Tang, Chen Tingzhang wrote *Waterwheel Poem*, and sketched its configuration. Liu Yuxi also gave an account of it in *Notes of Water Drawing Machine*.

The waterwheel was a huge standing wheel made of bamboo, setting up along the horizontal axis and surrounded by several semicircular bamboos or wooden buckets. It was installed by the waterside with the lower part of the standing wheel submerged in the river. When the wheel rotated continually via contact with the water and filled up the small surrounding buckets, the water was then poured into the fields via troughs.

The waterwheel largely prevailed in southwestern areas, and it adapted the local terrain and rushing currents with large changes. New developments were achieved in the Song and Yuan dynasties, such as the waterwheel and the high-and-low-waterwheel-complex driven by livestock, and the latter could draw water to a height of seven to eight zhang① (Chinese length measurement unit).

In short, in the Song Dynasty, the development of traditional irrigation tools basically reached the peak with no radical changes later on.

Words & Expressions

semi-arid/ˌsemi ˈærɪd/*adj.* 半干燥的
tillage/ˈtɪlɪdʒ/*n.* 耕作，耕种
solstice/ˈsɒlstɪs/*n.* 至，至日；至点
equinox/ˈekwɪnɒks；ˈiːkwɪnɒks/*n.* 春分；秋分；昼夜平分点
shadoof/ʃɑːˈduːf/*n.* 汲水吊杆，桔槔
windlass/ˈwɪndləs/*n.* 辘轳
chain pump/tʃeɪn pʌmp/*n.* [机] 链泵

Notes

1. Plow—harrow—level—press—hoe：犁—耙—平—压—锄。
2. Shadoof：桔槔俗称吊杆、称杆，古代汉族农用工具，是一种原始的汲水工具。商代在农业灌溉方面，开始采用桔槔。它是在一根竖立的架子上加上一根细长的杠杆，

---

① zhang，丈，中国古代长度单位，1 丈≈3.333m。

中间是支点，末端悬挂一个重物，前段悬挂水桶。一起一落，汲水可以省力。当人把水桶放入水中打满水以后，由于杠杆末端的重力作用，便能轻易把水提拉至所需处。桔槔早在春秋时期就已相当普遍，而且延续了几千年，是中国农村历代通用的旧式提水器具。这种简单的汲水工具虽简单，但它使劳动人民的劳动强度得以减轻。

3．Windlass：辘轳，汉族民间提水设施，流行于北方地区。由辘轳头、支架、井绳、水斗等部分构成。利用轮轴原理制成的井上汲水的起重装置。故曰："井上辘轳卧婴儿。"

4．ta tillage method：用人足或兽足来踩踏耕作田土，称为踏耕。

Reading Comprehension

I. Each piece of the following information is given in one of the paragraphs in the passage. Identify the paragraph from which the information is derived and put the corresponding number in the space provided.

(　　) 1. However, rainfall there remains very scarce, coupled with windy weather, often resulting in a poor harvest.

(　　) 2. The classic work *Arts for the People* was written during this time, which represented the highest level of world agronomy and marked the formation of China's intensive agricultural technology system.

(　　) 3. At the head of the tube a crank was fixed, and it controlled the up and down motion of the rope and bucket.

(　　) 4. With marvelous pumping capacity, it operated and transported water in a very convenient and efficient way.

(　　) 5. New developments were achieved in the Song and Yuan dynasties, such as the waterwheel and the high-and-low-waterwheel-complex driven by livestock, and the latter could draw water to a height of seven to eight zhang (Chinese length measurement unit).

II. Decide whether the statements are true (T) or false (F) according to the passage.

(　　) 1. China's Yellow River basin is a semi-arid, sub-humid region where the annual rainfall reaches only 600~750mm.

(　　) 2. After each plowing, a rake was used to break the earth, followed by the *mo* which further refined and flattened the topsoil to prevent water evaporation.

(　　) 3. Conversely, farming at the right time was also documented in the book, which led to predictable disasters.

(　　) 4. After the barrel was lowered to fill with water, it could be easily hoisted to the desired height due to the weight of the lever end.

(　　) 5. The waterwheel is another type of irrigation machine created in the Sui and Tang Dynasties.

III. Discuss the following questions.

1. What do you think of the waterwheel?
2. Talk about something you know about water Conservation on Dry Land in the Han and Wei Dynasties.

 灌溉工程
Famous Irrigation Projects

Dujiangyan

Located close to Chengdu, capital city of Sichuan Province, the Dujiangyan system was built by Qin, a kingdom during the Warring States Period in 256 BC. It is still in operation today by irrigating the farmland of over 668700 hectares along the Minjiang River. Before Dujiangyan was built, the areas on both sides of the Minjiang River were annually flooded. Li Bing, a governor of what is now Sichuan Province, was ordered to tame the river. After making field investigations, he found that the water from the surrounding mountains rushed down to join the slow-running water, thus swelling the river.

A dam could have been constructed to prevent floods. However, the river was then a very important waterway for military purposes and had to be kept open for vessels to ship troops and supplies. So genius was Li that instead of building a dam, he designed a unique system to divert running waters for irrigation, transportation, and flood control.

The system consisted of three major parts: Yuzui Dyke, Feisha Weir and Baoping Intake.

Yuzui Dyke is a V-shaped or fish-mouth dyke where the river water is diverted into two streams: the inner one and the outer one. The inner one is actually a deep and narrow channel carrying nearly 60% of the water into an irrigation system in dry seasons, while the outer one is the original waterway, so wide and shallow that much of the silt and sediment is flushed out during flood.

The dyke is a man-made islet constructed partially by strips of bamboo baskets which are filled with stones. These baskets are known as zhulong, and they are held in place by Macha, the wooden tripods. It took four years to complete the massive construction.

Feisha Weir or flying sand weir, which separates the inner stream from the outer one, is about 200 meters long, connecting the lower end of Yuzui Dyke and Baoping Intake. During flooding seasons, excessive water runs over the weir into the outer stream, draining out silt and sediment at the same time. When the water level is low, the weir ensures that there are enough water supplies in the inner stream for irrigation. Nowadays, a concrete weir has replaced the original one that was constructed with zhulong held in macha.

Baoping or Bottle-Neck Intake is a narrow entrance to a channel which transfers river water to farmlands. The 66-foot-wide entrance, which is dug by hand through part of a mountain, stretches into the Minjiang River. When building the intake, the rocks of the mountain were heated by fire and then cooled with water till they cracked and then were removed. People at that time spent eight years to complete the digging. Baoping Intake functions like a check gate, producing the whirlpool flows that carry away the excess water over Feisha Weir.

Since the system was completed, floods have been controlled and the area has turned into a land of abundance. On the east side of Dujiangyan, a temple was built in honor of Li Bing.

Ling Qu

Located in Xing'an County of Guilin in Guangxi Zhuang Autonomous Region, Ling Qu, or Magic Canal, runs through mountains, connecting the mid-low Yangtze River Valley with the Pearl River Delta of Guangdong Province, greatly promoting trade between these two regions. However, the canal was originally constructed for military use.

Around 221 BC, Ying Zheng defeated other kingdoms and became the first emperor of the Qin Dynasty, which covered a large area between the Yellow and Yangtze Rivers. Although he owned this vast territory, his ambition was not satisfied. He planned to extend his territory beyond the Wuling, a series of mountains that divide China into north and south parts, to Yue, a kingdom in what is now Guangdong Province. In 219 BC, he ordered to construct a canal to ship troops and provisions to the front lines.

Shi Lu was assigned to lead the task. However, the land across which the canal was to run was mountainous. Moreover, there were two rivers—Xiangjiang and Lijiang on the north and south of the mountains respectively. The former flowed north to join the great Yangtze River by way of Lake Dongting and the latter flowed south to join the Pearl River.

After plenty of field researches, Shi Lu had intake opened on the Xiangjiang at Xing'an County which diverted the river water into a man-made canal that has an almost horizontal bed with a slope that allows 30% discharge of the Xiangjiang waters. The canal runs for about five kilometers from the source of Lijiang, which was channelized for navigation along around 30 kilometers of its watercourse.

The significance of Ling Qu is the herringbone water-diverting system at the intake of the Xiangjiang River which includes a separation structure, a north canal (3.25 kilometers long) and a south canal (33.15 kilometers long).

The spear-head-like dyke allows 30% of Xiangjiang water to flow to the south canal. At the junction of the end of the dyke and the mouths of two canals are two 3.9-meter deep dams called Big Tianping and Small Tianping respectively. Such a complex system is used to raise the water level, ensuring sufficient water supply for the canal to run over the mountains. Besides, both dams are actually slopes. Therefore, the water can overflow during flooding seasons, avoiding the rupture of the dams.

Construction of Ling Qu was finished five years later in 214, which helped Ying Zheng conquer the Yue Kingdom. After that, the canal has been used for transportation and irrigation. As a comprehensive project making creative use of topographic factors and multiple hydraulic facilities, and as the first mountain-crossing canal in the world, it is renowned as a masterpiece of water engineering in human history.

Zhengguo Qu

Located in Shaanxi Province, Zhengguo Qu, or Chengkuo Canal, connected Jingshui River with Luoshui River, covering about 150 kilometers.

The canal was constructed to prevent the Hann Kingdom from being conquered by the Qin Kingdom, according to Sima Qian, one of China's greatest historians in the Han Dynasty. In order to protect itself from being attacked, Han Guo dispatched Zheng Guo, a water engineer, to persuade the ruler of Qin to construct a canal. The ruler of Han Guo hoped that the huge project would wear out Qin's labor and natural resources so that it couldn't build up forces to invade Hann. The project was launched around 246 BC and all went well before Hann's plot was uncovered when the canal was half finished.

The ruler of Qin was angry but did not have Zheng Guo killed, because Zheng convinced him that the Qin Kingdom could benefit greatly from the project in the long run, although it could not spare energy to invade Han Guo instantly as a result.

Zheng's words were proved to be true. Since the canal was put into service in 236 BC, the area was no longer struck by droughts and turned into a fertile land. With agricultural productivity improved considerably, Qin became richer and more powerful than ever before, which helped it to conquer all of the other kingdoms, including Hann.

The canal, which was named in honor of its designer Zheng Guo, was a grand project that was said to irrigate the farmland of 40000 qing, equal to 4 million mu, of which quite many used to have high soil salinity.

Without modern technology and equipment, Zheng Guo made an accurate topographical survey, ensuring the canal ran from high to low places. However, the Jingshui was a muddy river and the riverbed was gradually raised because of silt accumulation. Therefore, the canal had to be reconnected to the river with new intakes. By 95 BC, Zhengguo Qu was almost out of use, but it remained a perfect example for other canals to be built in this area.

Words & Expressions

1. hectare/'hekteə(r)/n. 公顷（等于 1 万平方米）
2. dam/dæm/n. 堤，坝
3. dyke/daɪk/n. 堤
4. tripod/'traɪpɒd/n. 三脚架

1. **Dujiangyan**：都江堰位于四川省成都市都江堰市城西，坐落在成都平原西部的岷江上，始建于秦昭王末年（约公元前256—前251），是蜀郡太守李冰父子在前人鳖灵开凿的基础上组织修建的大型水利工程，由分水鱼嘴、飞沙堰、宝瓶口等部分组成，2000多年来一直发挥着防洪灌溉的作用，使成都平原成为水旱从人、沃野千里的"天府之国"，至今灌区已达30余县市、面积近千万亩，是全世界迄今为止，年代最久、唯一留存、仍在一直使用、以无坝引水为特征的宏大水利工程，凝聚着中国古代劳动人民勤劳、勇敢、智慧的结晶。

2. **Ling Qu**：灵渠，古称秦凿渠、零渠、陡河、兴安运河、湘桂运河，是古代中国劳动人民创造的一项伟大工程。位于广西壮族自治区兴安县境内，于公元前214年凿成通航。灵渠流向由东向西，将兴安县东面的海洋河（湘江源头，流向由南向北）和兴安县西面的大溶江（漓江源头，流向由北向南）相连，是世界上最古老的运河之一，有着"世界古代水利建筑明珠"的美誉。2018年8月13日，灵渠等4个项目入选2018年（第五批）世界灌溉工程遗产名录。

3. **Zhengguo Qu**：郑国渠，于公元前246年（秦王政元年）由韩国水工郑国在秦国主持穿凿兴建，约十年后完工。是古代劳动人民修建的一项伟大工程，属于最早在关中建设的大型水利工程，位于今天的陕西省泾阳县西北25千米的泾河北岸。它西引泾水东注洛水，长达300余里。郑国渠在战国末年由秦国穿凿。2016年11月8日，郑国渠申遗成功，成为陕西省第一处世界灌溉工程遗产。

Reading Comprehension

I. Each piece of the following information is given in one of the paragraphs in the passage. Identify the paragraph from which the information is derived and put the corresponding number in the space provided.

(　　) 1. Li Bing, a governor of what is now Sichuan Province, was ordered to tame the river.

(　　) 2. These baskets are known as zhulong, and they are held in place by Macha, the wooden tripods.

(　　) 3. The 66-foot-wide entrance, which is dug by hand through part of a mountain, stretches into the Minjiang River.

(　　) 4. Around 221 BC, Ying Zheng defeated other kingdoms and became the first emperor of the Qin Dynasty, which covered a large area between the Yellow and Yangtze rivers.

(　　) 5. Construction of Ling Qu was finished five years later in 214, which helped Ying Zheng conquer the Yue Kingdom.

II. Decide whether the statements are true (T) or false (F) according to the passage.

(　　) 1. Since the canal was put into service in 236, the area was no longer struck by droughts and turned into a fertile land.

(　　) 2. Moreover, there were two rivers—Xiangjiang and Lijiang on the west and south of the mountains respectively.

(　　) 3. The 66-foot-wide entrance, which is dug by hand through part of a mountain, stretches into the Minjiang River.

(　　) 4. These baskets are known as *zhulong*, and they are held in place by *macha*, the wooden tripods.

(　　) 5. Before Dujiangyan was built, the areas on both sides of the Lijiang River were annually flooded.

III. Discuss the following questions.

1. Why do you think the Dujiangyan is a great project in ancient China?
2. Talk something about Zhengguo Qu?

第六章 传统农耕用具的技术革新

Chapter 6 Technological Innovations of Farming Implements

Text A 工具材料的创新 Innovation of Tool Materials

Stone, Wood, Bone (Mussel, Horn) and Others

The origin of agriculture can be traced back to the early and mid-Neolithic Age. Historians named the farming production in this period as slash and burn, e. g. a man used a polished rectangular rounded-blade stone ax to cut down wooded areas. After that, he waited for the stems to dry and the roots to rot, and then set fire to them. The remaining ashes were used as fertilizer. When sowing, women used short wooden bent-handled shovels or hoes to dig the soil. A pointed stick was also utilized during sowing. When crops were ripe, they were harvested by improved scrapers or simply hand-picked. They were then placed on a flat stone and with stone balls crushed or smashed to make them easier to eat. This was the prototype of the stone plate, stone rods and pestle mortar used today.

In all Neolithic agricultural sites across China, numerous tools were unearthed such as the stone ax, stone hoe, stone plow, stone knife, stone sickle, bone plow, mussel plow, bone sickle, mussel sickle, horn hoe, etc. These tools are part of the legacies of the Yangshao Culture, which dated back to 5000~7000 years ago. Generally, these tools were invented for soil preparation, harvesting and threshing.

Bronze Tools

The replacement of wood and stone tools by metal tools started with the use of bronze. The tools made of bronze, an alloy of copper and tin, were much sharper, harder and lighter.

Bronze jue was unearthed among the Shang Dynasty relics. In the Shang Dynasty, bronze was a precious metal, mainly used by nobles as tableware, ceremonial vessels and weapons with only a tiny part used for production tools. Farmers of the Zhou Dynasty attached great importance to intertillage. It was written in *Book of Songs* that jian was a bronze shovel and bo a bronze hoe. The bronze sickle appeared very early. Zhi mentioned in *Book of Songs* was the bronze claw sickle. Evolved from stone knives, it was used to cut ears of corn. The handled bronze sickle was called ai.

Iron Tools

Iron tools were first seen roughly from the late Western Zhou Dynasty to the Warring States Period. Early in the Spring and Autumn Period, Guan Zhong suggested casting farm tools with "inferior metal", whereas "superior metal" was to be utilized in weapon making. Iron softening technology was acquired during the Spring and Autumn Period. In particular, the emergence of malleable iron greatly improved iron productivity, which created very favorable conditions for the spread of iron farm tools. Iron farm tools were mainly categorized into jue, hoe, spade, shovel, sickle and plow.

All of the possibilities of social organization in ancient China were dramatically altered by the innovation of iron technology and the production of the iron plough, which began to appear late in the Spring and Autumn period. Prior to this time, plowshares had principally been made from wood. Wood ploughs could cut only looser types of soil and severely limited the area of land that qualified as arable. Moreover, wooden ploughs required a great deal of pressure to cut the soil to the degree they could. The farmer would need to grip the plough handle and use his foot to press the plough down as he pushed forward a step at a time, using the motion of his arms to clear a furrow as he went hopping along.

"The advent of the iron plough led directly to a second innovation: the ox-drawn plough. With the weight and sharpness of the iron plough, it was no longer necessary to exert more downward pressure to cut a furrow than provided by the arms as they gripped the plough 4 handle. This meant that the farmer's feet could be freed for walking. Of course, the iron of the plough was heavy and difficult to push. Over time, oxen, individually or in yoked teams, were increasingly employed to pull the plough. With the iron plough and ox power, lands that had once been too poor to be worked suddenly became worth opening up. Wastelands began to vanish, and in their stead settlements and walled towns sprang up. Many patrimonial estates of modest influence suddenly found themselves with a vastly increased tax base, anxious to attract the people of neighboring lands to their territories to provide labor for newly reclaimed fields. In some cases, buffer zones between states disappeared and became fertile country worth seizing. All of this contributed directly to the proliferation of armed struggle over the principal "means of production" land and labor.

Other innovations in this period contributed to these changes. In the past, farmers who wished to irrigate their crops had not had any method better than carrying jugs to and from water sources. During the Classical period, a device called a well sweep, which suspended a bucket at the end of a long pivoting lever, greatly eased small scale irrigation. In addition, the beginnings of a dense network of irrigation canals boosted yields, and allowed the cultivation of rice in areas where it had not been previously possible. A growing understanding of the importance of manure allowed the reduction in the amount of land that had to lie fallow, recovering nutrients, and also encouraged a burst in the planting of winter wheat, which made the double crop schedule feasible through a large portion of China.

Words & Expressions

mussel/'mʌsəl/*n.* 蚌；贻贝；淡菜
ax/æks/*n.* 斧头
hoe/həʊ/*n.* 锄头
plow/plaʊ/*n.* 犁
sickle/'sɪkl/*n.* 镰刀
thresh/θreʃ/*vt.* 打谷
bronze/brɒnz/*n.* 青铜；青铜色，古铜色；青铜艺术品；铜牌
alloy/'ælɔɪ/*n.* 合金；（与贵金属混合的）劣等金属
malleable/'mælɪəb(ə)l/*adj.* 有延展性的，易塑形的
yoke/jəʊk/*n.* 轭，牛轭；同轭的一对动物

Reading Comprehension

I. Each piece of the following information is given in one of the paragraphs in the passage. Identify the paragraph from which the information is derived and put the corresponding number in the space provided.

(　　) 1. In all Neolithic agricultural sites across China, numerous tools were unearthed such as the stone ax, stone hoe, stone plow, stone knife, stone sickle, bone plow, mussel plow, bone sickle, mussel sickle, horn hoe, etc.

(　　) 2. The origin of agriculture can be traced back to the early and mid-Neolithic Age.

(　　) 3. All of the possibilities of social organization in ancient China were dramatically altered by the innovation of iron technology and the production of the iron plough, which began to appear late in the Spring and Autumn period.

(　　) 4. In the past, farmers who wished to irrigate their crops had not had any method better than carrying jugs to and from water sources.

(　　) 5. The replacement of wood and stone tools by metal tools started with the use of bronze.

II. Decide whether the statements are true (T) or false (F) according to the passage.

(　　) 1. When crops were ripe, they were harvested by improved scrapers or simply hand-picked.

(　　) 2. These tools are part of the legacies of the Yangshao Culture, which dated back to 3000~5000 years ago.

(   ) 3. In the Qin Dynasty, bronze was a precious metal, mainly used by nobles as tableware, ceremonial vessels and weapons with only a tiny part used for production tools.

(   ) 4. Early in the Spring and Autumn Period, ShangYang suggested casting farm tools with "inferior metal", whereas "superior metal" was to be utilized in weapon making.

(   ) 5. During the Classical period, a device called a well sweep, which suspended a bucket at the end of a long pivoting lever, greatly eased small scale irrigation.

Language Focus

III. Complete the sentences with the correct form of the words in the table.

individually	emergence	origin	dramatically	classical
cultivation	replacement	In some cases	numerous	prior

1. The _____ of agriculture can be traced back to the early and mid-Neolithic Age.

2. In all Neolithic agricultural sites across China, _____ tools were unearthed such as the stone ax, stone hoe, stone plow, stone knife, stone sickle, bone plow, mussel plow, bone sickle, mussel sickle, horn hoe, etc.

3. Over time, oxen, _____ or in yoked teams, were increasingly employed to pull the plough.

4. _____, buffer zones between states disappeared and became fertile country worth seizing.

5. In addition, the beginnings of a dense network of irrigation canals boosted yields, and allowed the _____ of rice in areas where it had not been previously possible.

6. _____ to this time, plowshares had principally been made from wood.

7. In particular, the _____ of malleable iron greatly improved iron productivity, which created very favorable conditions for the spread of iron farm tools. Iron farm tools were mainly categorized into jue, hoe, spade, shovel, sickle and plow.

8. All of the possibilities of social organization in ancient China were _____ altered by the innovation of iron technology and the production of the iron plough, which began to appear late in the Spring and Autumn period.

9. During the _____ period, a device called a well sweep, which suspended a bucket at the end of a long pivoting lever, greatly eased small scale irrigation.

10. The _____ of wood and stone tools by metal tools started with the use of bronze.

IV. Match the sentences in Section A with the English translation in Section B.

Section A

1. 当庄稼成熟时，他们用改良的刮土机或简单的手工采摘来收获。

2. 这是今天使用的石板、石棒和杵臼的原型。

3. 此外，木制犁需要很大的压力才能将土壤切割到他们所能达到的程度。

4. 这意味着农民的脚可以自由行走。

5. 一般来说，这些工具是为土壤准备、收割和脱粒而发明的。

Section B

1. Moreover, wooden ploughs required a great deal of pressure to cut the soil to the degree they could.

2. This meant that the farmer's feet could be freed for walking.

3. When crops were ripe, they were harvested by improved scrapers or simply hand-picked.

4. This was the prototype of the stone plate, stone rods and pestle mortar used today.

5. Generally, these tools were invented for soil preparation, harvesting and threshing.

Ⅴ. Translate the paragraph into Chinese.

Iron tools were first seen roughly from the late Western Zhou Dynasty to the Warring States Period. Early in the Spring and Autumn Period, Guan Zhong suggested casting farm tools with "inferior metal", whereas "superior metal" was to be utilized in weapon making. Iron softening technology was acquired during the Spring and Autumn Period. In particular, the emergence of malleable iron greatly improved iron productivity, which created very favorable conditions for the spread of iron farm tools. Iron farm tools were mainly categorized into jue, hoe, spade, shovel, sickle and plow.

_____

_____

_____

_____

_____

Development

Ⅵ. Discuss the following questions.

1. What is the advantage of iron tools?

2. How many materials do you know to make tools?

种植工具的创新
Innovation of Planting Tools

Soil Preparation Tools

Shovel

The Shovel was a soil preparation tool directly inserted into the soil. It was rectangular, initially made of wood, then stone or a small quantity of bone. Shovels started to appear in the

Spring and Autumn Period and were popularized in the Warring States Period. In the Han Dynasty, the technical name of the shovel was included in *The Origin of Chinese Characters*.

In a number of northern sites of Jin and Yuan Dynasties, where shovels are often unearthed, the shape and size are quite similar to current iron spades.

Spade

Spade is another soil preparation tool that was directly inserted into the soil. Because it utilized set of metal blades, the abrasion was greatly reduced, and its digging potential was further enhanced. The tool appeared in the Shang Dynasty, spread widely in the Han Dynasty, and popularized until after the Northern and Southern Dynasties.

Bronze spades in the Shang and Zhou Dynasties were mostly in the form of the Chinese character "凹", whereas they took on flattened, arced or sharpened blades in the Spring and Autumn Period. In the Warring States period, it developed further and was made of iron.

Plow for Coupled Farming

Coupled plowing could be explained as follows: (a) shoulder-to-shoulder coupling a dual cooperation with one plowing while the other breaks the clods or dual cooperation with one behind the other; (b) a pair of leisi driven by dual laborers where they plow in opposite directions.

As coupled farming developed, the edge of leisi was gradually tilted forward and the handle curved backward. After continuous improvement, the plow came into being. The first plow might still carry the name leisi; however, it slowly differentiated from leisi with the dawn of cattle farming when it then got the name—plow.

During the early Eastern Han Dynasty, the government established a special management body and commissioned craftsmen for agricultural implement design, manufacture and improvement. According to *The Book of Han—Agriculture and Currency Annals*, all the planting and farming instruments had been made better and lighter. The unearthed physical objects and murals showed that the plow components in the early Eastern Han Dynasty included plow regulators, plow shaft, plow bottom, plow crossbar, mould board, etc. , among which the regulators controlled the depth and the width of the plowed field. The widespread use of the bridled bovine guide saved human labor, so in the Eastern Han, more often than not, the plow was operated by one man and two cows. The plowshare also varied a great deal with a very large plowshare used to create a furrow, a large one used on virgin soil and a small one to cultivate. In order to prolong its longevity, the plowshare tip was created. A variety of mould boards were designed to plow either for single or double labor. During this period plows were straight shaft plows.

Curved Shaft Plow

The curved shaft plow made its presence known in the south of the Yangtze River in the Tang Dynasty. It was further improved and more widely adopted in the Song Dynasty, marking the final evolution of the traditional Chinese plow. This plow had many distinct features. The yoke was not directly connected, but attached by a rope through its moving front panel or a hook. The plow shaft and the rope met below the cattle buttocks with the shaft shortened and transformed from straight to curve. With its frame weight further reduced, it became more flexible and comfortable and the plow could be pulled by only one cow. Its slenderness kept the plow stable during operation. Originally, this sort of plow was used in paddy farming.

Either leisi or jue was used to provide intermittent digging, whereas the plow worked continually with greater efficiency. The evolution from leisi to plow was the result of an accumulation of long-term agricultural production and labor experience, a revolution in digging and an extraordinary invention in the history of farming implements.

Planting Tools

Gourd seeding

Crop sowing underwent a long development from scatter to strip to machinery. A sowing tool called the gourd seeding was mentioned in Wang Zhen's *Agriculture Book* published in the Yuan Dynasty. Gourd looked like calabash. After ripening and drying of the fruit, its skin shrank to a very hard layer and was hollow inside. Then a hole was made in each end of the gourd. The top hole was fixed with a wooden handle, the bottom hole with a mouth from which grains dropped and in the upper part was made a small opening where seeds were filled with the bottom mouth placed against the furrow to be planted, a man held the gourd and filled with seeds and walked along the furrow with the seeds dropping from the mouth into the furrow below.

Seed plow

In the Qin and Han Dynasties, the seed plow was considered the most outstanding seeding implement. Based on previous strip sowing tools, the three-legged seed plow was an improvement made by Zhao Guo in the Han Dynasty. Wang Zhen's *Agriculture Book*, written in the Yuan Dynasty, Xu Guangqi's *Agricultural Policy Book* and Song Yingxing's *Ti'en kung k'ai-wu* (Tiangong Kaiwu) written in the Ming Dynasty, *A Collection of Ancient and Modern Books, A General Study of Timing*, written in the Qing Dynasty, as well as other well known agricultural books all recorded its usage and improvement. Compared with the two-legged seed plow frequently used in rural areas, there was no significant difference in structure. Its main frame included the plow frame, plow leg, plow shaft, seed hopper and guide tube. The seed plow was distinguished by its compactness, even seeding and a powerful efficiency that allowed it to sow wheat, soy bean, millet, sorghum and other crop seeds. It was an incredible creation of Chinese traditional agriculture. On the basis of which, multiple dung seed plow was created in the Yuan Dynasty, to furrow, seed, apply fertilizer, cap soil, etc.

Ta and dun

Ta and dun (踏和砘) were also seeding tools besides the seed plow. *Qi Min Yao Shu*, the Northern Wei Dynasty agricultural book, recorded ta, which indicated that it had existed long before the Northern and Southern Dynasties. Following the seed plow, ta flattened the furrow to cover the seeds. However, it didn't strike the soil hard enough, so sprouting was not ideal, and thus required manual pedaling. Later, in the Song and Yuan Dynasties, dun was created to compact the ground after sowing.

Seedling horse

A tool named the seedling horse used to pull and plant seedlings was created in the Song Dynasty, which saved farmers from bending their backs and waists. The two ends were higher, whereas the middle was concave, shaped like a small wooden boat. People rode on its back, transplanting rice seedlings with their hands, and sliding their feet into the mud to move the "horse". This not only kept it from sinking, but improved work efficiency and reduced labor intensity. When Su Shi, a great essayist and poet in the Northern Song, saw farmers employing such an apparatus in the city of Wuchang of present-day Hubei Province. He was inspired to write the *Seedling Horse Song* as a compliment to its ingenuity. Because it was easy to produce, and it was warmly welcomed by farmers in provinces of Hubei, Jiangxi, Guangdong and Zhejiang in the Song Dynasty.

shovel/'ʃʌv(ə)l/*n.* 铁铲
spade/speɪd/*n.* 锹
curved shaft plow *n.* 弯曲的轴犁
gourd/gʊəd；gɔːd/*n.* 葫芦，葫芦属植物
mural/'mjʊərəl/*n.* 壁画

Notes

dun：砘，播种后用石砘子把松土压实。

Reading Comprehension

I. Each piece of the following information is given in one of the paragraphs in the passage. Identify the paragraph from which the information is derived and put the corresponding number in the space provided.

(　　) 1. *Si* was a properly polished stone plate, with no handle, or with a small short handle only.

(　　) 2. Bronze spades in the Shang and Zhou Dynasties were mostly in the form of the Chinese character "凹", whereas they took on flattened, arced or sharpened blades in the Spring and Autumn Period.

(　　) 3. During the early Eastern Han Dynasty, the government established a special management body and commissioned craftsmen for agricultural implement design, manufacture and improvement.

(   ) 4. The evolution from *leisi* to plow was the result of an accumulation of long-term agricultural production and labor experience, a revolution in digging and an extraordinary invention in the history of farming implements.

(   ) 5. A tool named the seedling horse used to pull and plant seedlings was created in the Song Dynasty, which saved farmers from bending their backs and waists.

II. Decide whether the statements are true (T) or false (F) according to the passage.

(   ) 1. *Arts for the People*, the Tang Dynasty agricultural book, recorded *ta*, which indicated that it had existed long before the Northern and Southern Dynasties.

(   ) 2. Spade is another soil preparation tool that was directly inserted into the soil.

(   ) 3. In primitive society about 3000 to 4000 years ago, when humans were still living a savage life, farming tools were not yet specialized.

(   ) 4. In a number of northern sites of Qin and Yuan dynasties, where shovels are often unearthed, the shape and size are quite similar to current iron spades.

(   ) 5. The widespread use of the bridled bovine guide saved human labor, so in the Eastern Han, more often than not, the plow was operated by one man and two cows.

III. Discuss the following questions.

1. Suppose you were an ancient farmer which plant tool do you like best? Why?
2. Would you list some details of shovel?

收获割工具的创新
Text C　Innovation of Harvesting Tools

Harvesting Implements Sickle

In primitive times, the stone sickle, the mussel sickle and other harvesting implements were predominantly used. This changed during the Shang and Zhou Dynasties when the sickle underwent changes and developed into zhi and ai. Archaeological findings indicated that the iron sickle was in existence during the Warring States Period.

Undoubtedly, the most prominent harvest tool in history must be the push sickle. To use it, the farmer took the handle in hand, pushed it forward against the top layer of the field and quick harvest was done with the advantage of a much higher efficiency (several times of that of a hand sickle). The creation of the push sickle successfully transformed work patterns from non-continuous to continuous.

Scythe

In the Iron Age, the sickle became increasingly popular. Its type varied, some were large, such as bo (镈), a two-edged long-handled sickle in the pre-Qin period. In the Tang and Song Dynasties, bo evolved into a specialized tool for wheat harvesting—the wheat scythe. In Song and Yuan Dynasties, wheat harvester and wheat baskets were invented to work with the wheat scythe. Wheat harvester was a dustpan with two long mobile handles. Above it was installed a wheat scythe, protruding forward. The wheat scythe and wheat harvester were mobilized by a flexible manipulator fixed around the farmer's waist. The wheat in the distance was cut into wheat harvester. Wheat basket was tied by a rope to reaper's waist, which could move along, with wheels at the bottom. The wheat which fulfilled the harvester would be dumped into wheat basket later.

Threshing Tools

Flail

In the Spring and Autumn Period, the flail was mentioned in books such as *Discourses of the States*. If a farmer raised the long handle and spun it around the roller while striking the crops, the crop ears would fall off. This applied to wheat, millet, rice and other sorts of crop threshing. Flail scenes were presented in quite a few murals of the Dunhuang Mogao Grottoes.

Stone roller

The stone roller was used to grind grains under the force of the big columniform stone. To strengthen the effect, dual grinds or multiple grinds were adopted, and imprints were engraved around the surface of the rolling stone.

Threshing bed

In the reaping season, neither stone roller nor flail functioned well in the rain. Due to the fact that crops with grains couldn't be stored too long, farmers created thresh-while-harvest implements such as the barrel, slate and threshing bed. When the grains were raised and beaten against the box frames of the threshing bed, the grains would fall off with ease and convenience.

Dustpan, sieve

Detached by the threshing tools, the knocked-off grains always had quite a few hulls, leaves and other crumbs mixed in. In order to remove the debris, farmers invented lots of equipment, like the dustpan and sieve. No matter how differently they were shaped and constructed, there were basically two approaches to use them: (a) when the grains were winnowed by various tools, the weeds, hulls or dust would be blown off by natural winds; (b) the grains were filled in a certain container and after appropriate tossing and sifting, the debris could be removed. The two methods however, were not only laborious, but often restricted to natural conditions. For example, it was unlikely to thresh grains without wind. Later, with the aid of a winnower, working conditions and work efficiency were greatly improved.

Grain Winnower

The winnowing machine is also called the winnower or grain winnower. In the Western Han Dynasty, a seven-blade-fan was invented, which laid a solid foundation for the creation of the winnower. A poet in the Northern Song Mei Yaocheng wrote the poem *Tian Shan* in praise of the winnower. Wang Anshi, the famous litterateur in the Northern Song Dynasty, also chanted it in *Fifteen Verses for Farm Tools*. The winnowing machine was made completely of wood, with a fan-out exit at the rear and a drum-shaped wooden box at the front. The box contained a fan wheel made of four to six thin wooden planks. To use it, the farmer turns the crank of the axle so that the fan rotated. The top was fixed with a funnel to fill grains. Then the threshed grains dropped through the slot into the body of the fan. The airflow caused by the turning blades blew away lighter debris to the rear exit; whereas heavier grinds fell to the bottom and then out of the fan.

Grain Processing Implements

Pestle and mortar

The earliest method to process grain was to beat it with sticks and then to pestle. The first pestle was a piece of a crude wooden stick. The first mortar was a round pit cut in a tree trunk, thus called tree mortar, and then it developed into a special pit in the ground, covered with animal skins. The grains were dumped into the pit to pestle. This was the time when wood was broken as pestle and ground was dug as mortar. The pestle and mortar were invented during the rule of the Yellow Emperor, possibly even earlier. Later, farmers discovered that it was easier to remove the crop skin when ears were rubbed against a hard surface; consequently, millstone and stone frotton were invented.

Rice hulling hammer

Farmers easily became fatigued when working with wooden or stone pestle and it added great difficulty when processing large quantities of grain. Before the Western Han Dynasty, there had been one kind of pedaled pestle and mortar named the riding hammer. Based on the lever principle, a wooden pole was fit on a lever with one end fixed on a hammer head. A stone mortar was placed below the pole. When a man stepped on the other end of the pole, the hammer would head upward. As the foot was moved away, the hammer head dropped and knocked the grains. In the Han Dynasty, livestock and water-propelled rice hulling hammers were produced. In the Northern and Southern Dynasties, water-motored hulling hammers witnessed greater development. In order to cope with the quantity of the water flow, several hulling hammers could cooperate simultaneously, which was called the joined hulling hammer.

Water-propelled compound Huller

A tool named the rice huller was not significantly different from the mill. However, there's a flaw with the stone-made rice huller—it grounded off not only the shell or skin, but the grains itself. Later on, wood and soil hullers were invented and able to break the outer shells without damaging the inner grains.

The stone mill developed rapidly in the Western Han Dynasty. Since the stone were mostly rough and harsh, food particles were crushed coarsely. In the Eastern Han Dynasty, the stone became radial shaped, producing very fine powder. Besides this improvement, the hydraulic mill was invented and became commonplace in the Northern and Southern Dynasties. After further improvement, the water propelled rice hulling hammer was combined with the mill and grinder. This finally resulted in the creation of the compound waterwheel, which was designed to alternately grind flour, to take off the husk and to grind rice. These machines were manifestations of great achievements in science and technology during the Song and Yuan Dynasties, creating favorable conditions for agricultural development.

Words & Expressions

innovation/ˌɪnəˈveɪʃn/ *n*. 新事物，新方法；革新，创新
implements/ˈɪmplɪm(ə)nts/ *n*. 工具；器具
scythe/saɪð/ *n*. 长柄大镰刀；钐刀
threshing/ˈθreʃɪŋ/ *n*. 脱粒；打谷
flail/fleɪl/ *v*. 用连枷打（谷）；*n*. 连枷（旧时打谷物用的工具）
roller/ˈrəʊlə(r)/ *n*. 碾子
dustpan/ˈdʌstpæn/ *n*. 簸箕
sieve/sɪv/ *n*. （粮食）筛子
winnower/ˈwɪnəʊə(r)/ *n*. 风车；扬谷器
pestle/ˈpesl/ *n*. 杵，碾槌；*v*. 用杵捣，用槌磨
mortar/ˈmɔːtə(r)/ *n*. 研钵，臼
rice hulling hammer *n*. 稻谷脱壳锤
propelled/prəˈpeld/ *adj*. 推进的

Reading Comprehension

I. Each piece of the following information is given in one of the paragraphs in the passage. Identify the paragraph from which the information is derived and put the corresponding number in the space provided.

(　　) 1. In the Tang and Song Dynasties, bo evolved into a specialized tool for wheat harvesting—the wheat scythe.

(　　) 2. When the grains were raised and beaten against the box frames of the threshing bed, the grains would fall off with ease and convenience.

(　　) 3. Before the Western Han Dynasty, there had been one kind of pedaled pestle and mortar named the riding hammer.

(　　) 4. These machines were manifestations of great achievements in science and technology during the Song and Yuan Dynasties, creating favorable conditions for agricultural development.

(　　) 5. The airflow caused by the turning blades blew away lighter debris to the rear exit; whereas heavier grinds fell to the bottom and then out of the fan.

II. Decide whether the statements are true (T) or false (F) according to the passage.

(　　) 1. The earliest method to process grain was to beat it with sticks and then to pestle.

(　　) 2. In the East Han Dynasty, a seven-blade-fan was invented, which laid a solid foundation for the creation of the winnower.

(　　) 3. In the Spring and Autumn Period, the flail was mentioned in books such as *Discourses of the States*.

(　　) 4. This changed during the Han Dynasties when the sickle underwent changes and developed into zhi and ai. Archaeological findings indicated that the iron sickle was in existence during the Warring States Period.

(　　) 5. When the grains were raised and beaten against the box frames of the threshing bed, the grains would fall off with ease and convenience.

Development

III. Discuss the following questions.

1. Choose one of the threshing tools to describe.
2. Choose one of the grain processing implements to describe.

第七章　农耕时令——二十四节气

Chapter 7　Farming Time—The 24 Solar Terms

Text A 二十四节气
The 24 Solar Terms

The "24 solar terms" originated and was firstly used in China. It was created thousands years ago on the basis of practical needs of agriculture. It's determined by the changes of the sun's position in the zodiac throughout the year, with two segments each month. Ancient Chinese people used it to guide agriculture production, special climate signs such as flood and drought, cultural ceremonies, family gatherings and even healthy living tips. Nowadays the "24 solar terms" is still being used by farmers and followed in many other parts of the world apart from China. As the fifth great invention of china, it was inscribed by the UNESCO on the *Representative List of the Intangible Cultural Heritage of Humanity* in 2016.

China is a great country of agriculture and has a long history of agricultural production for over 5000 years. Agricultural production is highly subjected to the seasonal changes. Therefore, at the initial stage of agricultural development, Chinese people began to observe the regularity and variations of climate and explore the seasonal rules in the agricultural production. At the beginning, Chinese laborers measured time by observing the movements of the sun, moon and stars.

It was said that during the Spring and Autumn Period the ancient people used the bamboo pole to measure the variation of time. They saw the different length of the shadow of the bamboo pole at the same time in a day. They carefully observed the differences and found that there was one day of a year on the noon of which the shadow of the bamboo was the shortest and the daytime was the longest and another day on the noon of which the shadow of the bamboo was the longest and the daytime was the shortest. They gave them the names of the Summer Solstice and the Winter Solstice. So the Summer Solstice and the Winter Solstice were firstly set. Then Chinese ancestors observed that there were two days in spring and autumn of a year on which the length of the day and that of the night are the same, and they defined them Spring Equinox and the Autumn Equinox. Later added were the Beginning of Spring, the Beginning of Summer, the Beginning of Autumn and the Beginning of Winter. During the early years of Western Han Dynasty, the complete 24 solar terms came into being. The following is the 24 solar terms in Chinese characters and ping yin.

The Song of the 24 Solar Terms

In order to remember the 24 solar terms, Chinese people read the 24 solar terms together with one character of each solar term omitted, so the song of the 24 solar terms was formed. To be short, the song is read as follows:

chūn yǔ jīng chūn qīng gǔ tiān xià mǎn máng xià shǔ xiāng lián
春雨惊春清谷天，夏满芒夏暑相连，

qiū chù lù qiū hán shuāng jiàng dōng xuě xuě dōng xiǎo dà hán
秋处露秋寒霜降，冬雪雪冬小大寒。

Table 7.1 shows the correspondent of abbreviation form in the song of the 24 solar terms.

Table 7.1 Terms and Full Title and Their Meanings

Abbreviation	Full Title	English	Meaning
春	立春	Beginning of Spring	Spring begins
雨	雨水	Rain Water	Rainfall begins and gradually increases
惊	惊蛰	Waking of Insects	Hibernating animals come to sense
春	春分	Spring Equinox	Day and night in spring are equally long
清	清明	Pure Brightness	Warm and bright in northern China with grass and trees flourishing
谷天	谷雨	Grain Rain	Rainfall is plentiful and comes at the right time, which helps crops to grow nicely
夏	立夏	Beginning of Summer	Summer begins
满	小满	Slight Fullness	Wheat begins to fill out
芒	芒种	Grain in Ear	Wheat grows ripe
夏	夏至	Summer Solstice	Summer truly begins and daytime is the longest and nighttime is the shortest
暑	小暑	Slight Heat	The weather starts to get hot
	大暑	Great Heat	The hottest time the year has arrived
秋	立秋	Beginning of Autumn	Autumn begins
处	处暑	Limit of Heat	The heat hides and the hot summer days come to end
露	白露	White Dew	Dew curdles
秋	秋分	Autumn Equinox	The day and night in autumn are equal in length
寒	寒露	Cold Dew	The dew is cold, which will soon form ice
霜降	霜降	Frost's Descent	The weather is increasingly colder and the frosts descent

（续）

Abbreviation	Full Title	English	Meaning
冬	立冬	Beginning of Winter	Winter begins
雪	小雪	Slight Snow	It begins to snow
雪	大雪	Heavy Snow	Snowfall increases and snow may accumulate
冬	冬至	Winter Solstice	Cold winter arrives in northern hemisphere and in this date daytime is the shortest and nighttime is the longest
小	小寒	Slight Cold	It is rather cold
大寒	大寒	Great Cold	The coldest time of the year has arrived

The Significance of the 24 Solar Terms

The 24 solar terms can be divided into the following categories. Eight of them indicate the changing of the four seasons. They include the Beginning of Spring, Summer, Autumn and Winter, the Summer and Autumn Equinoxes, and the Spring and Autumn Equinoxes.

Of them the Beginning of Spring, the Beginning of Summer, the Beginning of Autumn, and the Beginning of Winter mark the beginning of the seasons.

The Beginning of Summer indicates that summer begins. After the Beginning of Summer, the temperature will increase greatly and the weather will change quickly.

The Spring Equinox, Autumn Equinox, the Summer Solstice, and the Winter Solstice indicate the turning points of the seasons. The spring and autumn equinox refer to the point where the length of the days and nights are the same.

Another seven are about weather phenomena, including Rain Water, Grain Rain, White Dew, Cold Dew, Frost's Descent, Slight Snow and Heavy Snow. They show the trend of the weather in the following couple of days.

In addition to these, another five terms demonstrate changes in temperature, including: Slight Heat, Great Heat, End of Heat, Slight Cold, and Great Cold.

The rest of four are terms mirroring phenological phenomena, including Waking of Insects, Pure Brightness, Slight Fullness and Grain in Ear. Especially, Slight Fullness and Grain in Ear show the phenomenon of crop growth and promise the proper time for harvesting.

So we can say the 24 solar terms reflect changes that occur through the seasons, guide farm work, and influence every aspect of life for hundreds of thousands of families.

Words & Expressions

intangible/ɪn'tændʒəbl/*adj.* 不可捉摸的，难以确定的；（资产，利益）无形的
solstice/'sɒlstɪs/*n.* 至，至日；至点
equinox/'ekwɪnɒks；'iːkwɪnɒks/*n.* 春分；秋分；昼夜平分点
zodiac/'zəʊdiæk/*n.* 黄道带，十二宫图；属相，星座
flourish/'flʌrɪʃ/*v.* 繁荣，昌盛；挥动；（植物或动物）长势好，茁壮成长

Notes

The 24 solar terms：二十四节气是中国人认知一年中时令、气候、物候等方面变化形成的知识体系和社会实践，是中国传统历法体系及其相关实践活动的重要组成部分。在国际气象界，这一时间知识体系被誉为中国的第五大发明。二十四节气是将黄道假定为个大圆，太阳从黄经0°起，沿黄经每运行15°所经历的时日称为"一个节气"。每年运行360°，共经历二十四个节气。由于二十四节气主要反映太阳的周年视运动，所以在公历中，它们的日期是相对固定的，上半年的节气在6日、21日，下半年的节气在8日、23日左右，前后不差12日。其中，每月第个节气为"节气"，即立春、惊蛰、清明等12个节气；每月的第二个节气为"中气"，即雨水、春分、谷雨等12个节气。"节气"和"中气"交替出现，各历时约15天，现在人们已经把"节气"和"中气"统称为"节气"。

Reading Comprehension

I. Each piece of the following information is given in one of the paragraphs in the passage. Identify the paragraph from which the information is derived and put the corresponding number in the space provided.

( ) 1. Ancient Chinese people used it to guide agriculture production, special climate signs such as flood and drought, cultural ceremonies, family gatherings and even healthy living tips.

( ) 2. During the early years of Western Han Dynasty, the complete twenty four solar terms came into being.

( ) 3. Of them the Beginning of Spring, the Beginning of Summer, the Beginning of Autumn, and the Beginning of Winter mark the Beginning of the seasons.

( ) 4. The Spring Equinox, Autumn Equinox, the Summer Solstice, and the Winter Solstice indicate the turning points of the seasons.

( ) 5. Another seven are about weather phenomena, including Rain Water, Grain Rain, White Dew, Cold Dew, Frost's Descent, Slight Snow and Heavy Snow.

II. Decide whether the statements are true (T) or false (F) according to the passage.

( ) 1. Nowadays the "24 solar terms" is still being used by farmers and followed only in China.

( ) 2. It was said that during Qin Dynasty, the ancient people used the bamboo pole to measure the variation of time.

( ) 3. Another five terms demonstrate changes in temperature, including: Slight Heat, Great Heat, End of Heat, Slight Cold, and Great Cold.

( ) 4. The rest of four are terms mirroring phenological phenomena, including Waking of Insects, Pure Brightness, Slight Fullness and Grain in Ear.

( ) 5. We can say the 24 solar terms reflect changes that occur through the seasons, guide farm work, and influence every aspect of life for hundreds of thousands of families.

Language Focus

III. Complete the sentences with the correct form of the words in the table.

solstice	equinox	indicate	phenomenon	accumulate
intangible	regular	correspondent	zodiac	flourish

1. It's determined by the changes of the sun's position in the _____ throughout the year, with two segments each month.

2. As the fifth great invention of China, it was inscribed by the UNESCO on the Representative List of the _____ Cultural Heritage of Humanity in 2016.

3. Therefore, at the initial stage of agricultural development, Chinese people began to observe the _____ and variations of climate and explore the seasonal rules in the agricultural production.

4. They gave them the names of the Summer _____ and the Winter Solstice.

5. Then Chinese ancestors observed that there were two days in spring and autumn of a year on which the length of the day and that of the night are the same, and they defined them Spring _____ and the Autumn Equinox.

6. The next table shows the _____ of abbreviation form in the song of the 24 solar terms.

7. Warm and bright in northern China with grass and trees _____.

8. Snowfall increases and snow may _____.

9. The Spring Equinox, Autumn Equinox, the Summer Solstice, and the Winter Solstice _____ the turning points of the seasons.

10. Another seven are about weather _____, including Rain Water, Grain Rain, White Dew, Cold Dew, Frost's Descent, Slight Snow and Heavy Snow.

IV. Match the sentences in Section A with the English translation in Section B

Section A

1. 根据长期的观测发现，在夏天的某一天，正午竹影最短，在冬天的某一天，正午竹影最长，于是便确立了最早的两个节气——夏至和冬至。

2. 其中立春、立夏、立秋和立冬标志着季节的更替。

3. 二十四节气起源于中国。早在几千年以前，人们根据农耕的生产实践创造出二十四节气。

4. 反映气温变化的有小暑、大暑、处暑、小寒和大寒。

5. 二十四节气被誉为中国的第五大发明，2016年联合国教育、科学、文化组织列入世界非物质遗产名录。

Section B

1. The "24 solar terms" originated and was firstly used in China. It was created thousands

years ago on the basis of practical needs of agriculture.

2. As the fifth great invention of china, it was inscribed by the UNESCO on the Representative List of the Intangible Cultural Heritage of Humanity in 2016.

3. They carefully observed the differences and found that there was one day of a year on the noon of which the shadow of the bamboo was the shortest and another day on the noon of which the shadow of the bamboo was the longest. They gave them the names of the Summer Solstice and the Winter Solstice.

4. Of them the Beginning of Spring, the Beginning of Summer, the Beginning of Autumn, and the Beginning of Winter mark the Beginning of the seasons.

5. Another five terms demonstrate changes in temperature, including: Slight Heat, Great Heat, End of Heat, Slight Cold, and Great Cold.

V. Translate the paragraph into Chinese.

The ancient Chinese divided the sun's annual circular motion into 24 segments. Each segment was called a specific "solar term". The element of 24 solar terms originated in the Yellow River reaches of China. The criteria for its formulation were developed through the observation of changes of seasons, astronomy and other natural phenomena in this region and has been progressively applied nationwide. It starts from the Beginning of Spring and ends with the Great Cold, moving in cycles. The element has been transmitted from generation to generation and used traditionally as a timeframe to direct production and daily routines. It remains of particular importance to farmers for guiding their practices. Having been integrated into the Gregorian calendar, it is used widely by communities and shared by many ethnic groups in China. Some rituals and festivities in China are closely associated with the solar terms for example, the First Frost Festival of the Zhuang People and the Ritual for the Beginning of Spring in Jiuhua.

VI. Discuss the following questions.

1. How is agricultural production highly subjected to the seasonal changes?

2. How were the Summer Solstice and the Winter Solstice named?

Text B 二十四节气解说（1）
The Exact Interpretations of the 24 Solar Terms (1)

Beginning of Spring is the first solar term in the whole year. Li means beginning, and Beginning of Spring opens a spring world and shows the world comes to itself after a whole winter's rest and preservation. According to Chinese traditional medicine, the health care in spring should focuses on following the nature and protecting the yangqi in body. Besides, spring has the nature of wood, and it is most closely connected with liver, so in spring taking care of the liver is the most important in terms of Chinese traditional medicine theory.

Rain Water is the period of rainfall in the whole year. In Chinese traditional medicine theory, the period of the Rain Water plays the important role in caring spleen and stomach. The harmony of spleen and stomach can improve and regulate the physical metabolism. In Chinese traditional medicine, the root of human health is yuanqi (Original qi), and the root of yuanqi in human body is spleen and stomach. The poor spleen and stomach are the predominant causes of different diseases. So in the aspect of eating, people had better not eat too much oily and fat food but the red Chinese date, lotus, leek, spinach, orange, honey and sugarcane.

Waking of Insects is the period that the various dormant insects are awaken in earth by the warming weather and spring thundering. During this solar term, people with different corporeities had better take care of themselves in the aspects of spirit, environment and dining.

Spring Equinox means that the yin and yang occupy half the day respectively, and the whole day is evenly comprised of daytime and nighttime, and the temperature is even between warmth and coldness. Spring Equinox goes halves with daytime and nighttime as well as coldness and warmth, so people should take notice of the yin yang balance in the body in the aspect of health care. Today's most threatening cancers and disease of heart-blood vessel in the world are closely connected with the internal material exchange imbalance. So the experts of Chinese traditional medicine emphasize the method that regulating yin yang, the body deficiencies should be added with relevant supply and the extras should be let out.

Pure Brightness means the clearness of heaven and earth. An important traditional festival, Tomb-Sweeping Day, is observed in this period. It is also the peak time of high blood pressure. the exterior bad stimulation and long-time mental intension, anxiety and affliction may worsen the symptom of hypertension. In the mean time, it is the best time to select the soft and mild shadowboxing as the way of daily physical exercise.

Grain Rain is a turning point and since then the temperature ascends again. From then on, the rainfall increases, and the sufficient rainfall can well irrgate the newly-plant seedings. During this time, the weather is mainly sunny and warm; but the temperature of the dawn and dark may be unstable, so people had better keep the comfort of body to avoid the unwanted sufferings.

Beginning of Summer means the spring goes away and summer comes. Traditionally, people take Beginning of Summer as temperature rising. The weather is featured with the hottness and thunder storm. The crop rapidly grows up in this solar term. After Beginning of Summer, the weather becomes hotter and hotter, and the forest becomes thick. But this period is good for the heart because the heart shows its most vivid power and functionality in this term. Therefore, during this time, it is better to keep pleasure and ease both in mental and physical. Grief and rapture are harmful to health care.

Slight Fullness is the period that summer crops like barley and wintry wheat nearly but not absolutely mature, hence it is called xiaoman. During this period, people had better follow the health-care value of precaution of illness. The dining should be mainly inclined to the vegetables.

Grain in Ear belongs to a hot and moist period, especially in the midstream and downstream regions of Yangtze River. During this period, people easily get exhausted and weak. So people should focus on the spiritual care and keep easy and happy. In dining, people had better eat the food benefiting the body for facing the heat.

Summer Solstice used to be the oldest festival in ancient China. In health care, this period is the peak time of yangqi in accordance with Chinese traditional medicine theory. The health care had better focus on sleeping later and getting up earlier due to the superiority of yangqi in summer, so it is very important to arrange well the lunch break, and taking a warm shower everyday is also recommended. The daily exercise should not be too extreme or acute lest the overabundance of sweat. In dining, eating more sour food is good for the superficiality and eating some salty food is good for heart, but people had better not eat too much fat food.

Slight Heat is the high time of rainfall and also the time that the fire-worms are flying outdoors. Due to the heat, people may feel uneasy and exhausted so the traditional medicine experts may suggest that it is the time for caring the heart, over joy and grief are both harmful to heart. In dining, people had better pay more attention to the abstinence and take notice of individual's sanitation. Five tastes (sour, bitter, sweet, hot and salty) should be balanced in normal life, and never over drink too much alcohol and surfeit.

Great Heat is the hottest period all the year. The climatic feature is heat, torridity and raininess. So the heat and moisture both undermine the health. Therefore, people had better properly arrange the work and take notice of work and rest. People also should pay more attention to lowering the indoor temperature and having sufficient sleeping. In dining, people had better have some herbal porridge to preserve the body and avoid the disease and senescence. Per day, drinking a cup of fresh and cool boiled water is recommended.

spleen/spli:n/*n.* 脾脏

metabolism/mə'tæbəlɪzəm/*n.* 新陈代谢

spinach/'spɪnɪtʃ/n. 菠菜
dormant/'dɔːmənt/adj. 休眠的；静止的；睡眠状态的；隐匿的
corporeity/,kɔːpə'riːɪti/n. 身体，体质
shadowboxing/'ʃædəʊ,bɒksɪŋ/n. 太极拳
rapture/'ræptʃə(r)/n. 兴高采烈
superficiality/,suːpə,fɪʃi'æləti; ,sjuːpə,fɪʃi'æləti/n. 浅薄，肤浅；表面性的事物
abstinence/'æbstɪnəns/n. 节制；节欲；戒酒；禁食
sanitation/,sænɪ'teɪʃn/n. 公共卫生，环境卫生；下水道设施，卫生设备
torridity/'tɒrɪdɪti/n. 炎热
senescence/sɪ'nesns/n. 衰老

1. In Chinese traditional medicine theory, the period of the Rain Water plays the important role in caring spleen and stomach.
雨水时节养生，健脾养胃是关键。脾胃是后天之本，人体中气血皆是由此而来。

2. So in the aspect of eating, people had better not eat too much oily and fat food but the red Chinese date, lotus, leek, spinach, orange, honey and sugarcane.
在雨水时节吃红枣桂圆莲子百合粥比较好，它里面所含有的大枣、桂圆、莲子等食材本身就有滋补的功效，而把它们放在一起煮粥吃就会有生津润燥，安神养血的作用，这对于那些心脾两亏和气血不足的人群来说有非常好的改善作用。
雨水时节应少吃油腻食物，而食用大枣、莲子、韭菜、菠菜及橙子、蜂蜜、甘蔗等。

Reading Comprehension

I. Each piece of the following information is given in one of the paragraphs in the passage. Identify the paragraph from which the information is derived and put the corresponding number in the space provided.

(   ) 1. Rain Water is the period of rainfall in the whole year.
(   ) 2. Waking of Insects is the period that the various dormient insects are awaken in earth by the warming weather and spring thundering.
(   ) 3. The weather is featured with the hottness and thunder storm.
(   ) 4. Grain in Ear belongs to a hot and moist period, especially in the midstream and downstream regions of Yangtze River.
(   ) 5. Slight Heat is the high time of rainfall and also the time that the fire-worms are flying outdoors.

II. Decide whether the statements are true (T) or false (F) according to the passage.

(   ) 1. In summer taking care of the liver is the most important in terms of Chinese traditional medicine theory.

(　　) 2. Spring Equinox goes halves with daytime and nighttime as well as coldness and warmth.

(　　) 3. From Grain Rain on, the rainfall increases, and the sufficient rainfall can well irrgate the newly-plant seedings.

(　　) 4. Traditionally, people take Summer Solstice as temperature rising.

(　　) 5. Great Heat is the hottest period all the year. The climatic feature is heat, torridity and raininess.

Development

Ⅲ. Discuss the following questions.

1. Why Slight Fullness is called xiaoman?

2. Why is it significant to arrange well the lunch break, and taking a warm shower in summer every day?

Text C 二十四节气解说（2）
The Exact Interpretations of the 24 Solar Terms (2)

Beginning of Autumn is the time that the cool wind comes and the heat leaves, and also means the coming of autumn. It is the transfer period from the hot to the cool. It is the time that yangqi is more and more inferior and yinqi gets more and more superiority, so in Chinese traditional medicine theory, it is the transition of yin and yang in metabolism of human body. The harvest of crop is also during this period. Traditionally, Chinese peopel have the sense of sorrow in autumn. In autumn, it is the time to take care of the lung, the poor qi of lung may descend its acceptance of the bad stimulation, so it is easy for people to be grieved. In autumn, it is also the golden time to take diversity of physical exercises to keep the function of lung.

Limit of Heat means the heat is past. This is a transition time from hotness to coolness. So the weather is not so stable, some days may be quite hot, some may be lower than expectations. In folk China, people usually describe it into a fall tiger, which is crueler than tiger. During this time, the sufficient rest and high-quality sleeping is more considerably important. Experts of Chinese traditional medicine emphasize the importance of sleeping at night and noon. In dining, it is better to have some purely-warm food.

White Dew is a typical phase of autumn. From this day on, the dews appear and become more and more outdoors. White Dew is the real beginning of the cool autumn. During this time, people should pay more attentions to some chronics like nasal cavity disease, asthma and bronchia disease. In dining, the food should be easily digestive and rich in vitamin. What is more, the dryness is the typical climatic feature of autumn, and it is called Autumn Dryness in folk society, so it is time to eat more food rich in vitamin, and also suitable for taking some traditional medicines good for getting rid of sputum.

Autumn Solistice means that the sunshine nearly directly shines the equator. And the whole day is equally comprised of half daytime and half nighttime. As the phase that the nighttime and daytime go halves, people had better take care of the balance of Yin and Yang in human body. In dining, people should focus on the adding the deficiency and cutting the extras to show the balance of interiority of human body, and it also shows the principle of warming the cold and cooling the warm.

Cold Dew means the cold air will be frozen into dews. In Chinese traditional medicine, experts focus on the health care of caring Yang in spring and summer as well as caring yin in autumn and winter. Anciently, the autumn used to be called Golden Autumn, and in the five internal organs of human body, the lung air corresponds to the air of golden autumn. If the care is insufficient and improper, people may have a series of seasonal symptoms like skin desiccation, nose and pharynx dryness, so it is the time to eat more food like dairy products, honey, sticky rice and sesame. And more meat should be included as well to strengthen the body.

Frost's Descent means the weather becomes cold and frost appears. This phase is in the dominancy of Earth, one of five elements, and in traditional medicine, the food should be simple and less fat, but people had better pay more attention to add more vigor to care about the stomach. Chestnut is good for stomach, kidney and spleen, so it is the ideal seasonal food in winter. Additionally the vegetables like spinach and Chinese watermelon covered by frost is also quite delicious. The grape in winter is very sweet as well. All of them are all the good selections for eating. Ancient people in winter used to eat the mutton and rabbit meat, and it is mostly reasonable.

Beginning of Winter means the coming of winter. Due to the vastness of Chinese territory, except the south coastal area without the effect of winter and Qinghai-Tibet Altiplano without the effect of summer, the other areas do not enter winter by the arrival of Beginning of Winter. The northeast region enters winter as early as September. In the late period of October, Beijing has been influenced vividly by winter. While in the region of Yangtze River, the influence of winter becomes stronger and stronger around the phase called Slight Snow. In ancient times, Beginning of winter was an important festival. On this day, the emperor followed by a group of officials would come to the suburb of capital to hold a large sacrifice to the heaven. Nowadays, people still make some specials such as eating jiaozi, winter swimming, eating the mutton and beef as well as eating carrots.

During Slight Snow the southern China gradually enters winter, and the winter landscape has been vivid. In dining, people had better take some food good for kidney like mutton, beef, bone soup, stewed meat with chestnut, and chicken. The fruits include cashew and chestnut etc. are also good for health. Additionally, eating more stewed food and black food like Jew's-ear, black gingili and black beans. After the Slight Snow, the temperature rapidly falls, and the weather is inclined to the dryness, and it is the better time to make some preserved on salted food. In China, people easily find the preserved hams in each family. In southern China, local people also have the custom of eating ciba. During this phase, the weather is cold and gloomy, which surely negatively influences the mood of people. So it is a dangerous time for those people who suffer melancholia. It is more important to keep a good mood during this time.

Heavy Snow means the times of snowfalls increase, and the quantity of snow becomes large. In ancient China, this phase can be divided into three sections, that is, gulls and cuckoo stop singing, the tiger begins searching for partner and Chinese small iris comes out. In Chinese traditional medicine, the great cold is the perfect time to care about the health. Additionally, during this period, the rainfall is less than anytime else, and the weather is quite dry. So it is better time to eat more fruits and vegetables like oranges, apples and winter Chinese dates.

Winter Solstice has the shortest daytime and longest nighttime in north hemisphere all the year around. On this day, in many regions of northern China, people have the custom of eating jiaozi, and southern China, people eat Tangyuan. Winter solstice is a good time for health care, in accordance to the saying of Chinese traditional medicine that the qi appears from Winter Solstice. From this day on, the life declines from vigor. So eating some selected food can keep the vigor of life, which is also good for longevity. The dining during this phase should be various, but the too fat and salty food should not be eaten too much.

In China, Slight Cold means the coldest time comes. According to the climatic data of China, the Slight Cold is the phase with the lowest temperature. In folk society, people also say the all-year coldness appears in San Jiu, which roughly lasts from January 9th to January 17th. In Chinese traditional medicine theory, the coldest time is also the highest time of yinqi. In dining, people had better eating more warm and hot food to prevent the coldness. So, it is the best time to enjoy the hotpot and chaffy dish.

Great Cold is the last solar term and it means the most extreme cold time appears in the whole year. Due to the transition of Great Cold and Beginning of Spring, in dining, people also follow this change. People had better not eat too much but select some easily-digesting food to suit the Beginning of Spring. In Foshan, Guangdong Province, the local has the custom of braising the sticky rice, which has the vivid role in preventing coldness. In Anqing, Anhui Province, the local has the custom of frying chunjuan. All in all, the end of 24 solar terms also means a new beginning next year.

Words & Expressions

superiority/suːˌpɪəriˈɒrəti/*n*. 优越，优势；优越感，骄傲自大

dew/djuː/*n*. 露水；（似露珠的）小水珠

nasal/'neɪzl/*adj.* 鼻的；（嗓音，话语）带鼻音的；（语音）鼻音的
cavity/'kævəti/*n.* 洞，腔；(牙齿的) 龋洞
asthma/'æsmə/*n.* 哮喘，气喘
bronchia/'brɒŋkɪə/*n.* 支气管
sputum/'spjuːtəm/*n.* 唾液
desiccation/ˌdesɪ'keɪʃn/*n.* 干燥
pharynx/'færɪŋks/*n.* [解剖] 咽，咽喉
cashew/'kæʃuː/*n.* 腰果
gingili/'dʒɪndʒɪli/*n.* 芝麻；芝麻油
melancholia/ˌmelən'kəʊliə/*n.* 忧郁症
cuckoo/'kʊkuː/*n.* 布谷鸟；杜鹃鸟
iris/'aɪrɪs/*adj.* 鸢尾属植物的

Notes

1. Beginning of Autumn: 立秋不仅预示着炎热的夏天即将过去，秋天即将来临，也表示草木开始结果孕子，收获季节到了。

2. Autumn Solistice: 秋分时太阳直射赤道，昼夜均分，全球无昼极夜现象。秋分之后白天逐渐变短，黑夜变长，气温逐日下降，逐渐步入深秋季节。

3. Beginning of Winter: 立冬不仅意味着冬季的开始，还有万物收藏、规避寒冷的意思。

Reading Comprehension

Ⅰ. Each piece of the following information is given in one of the paragraphs in the passage. Identify the paragraph from which the information is derived and put the corresponding number in the space provided.

(   ) 1. Beginning of Autumn is the transfer period from the hot to the cool. It is the time that yangqi is more and more inferior and yinqi gets more and more superiority.

(   ) 2. Cold Dew means the cold air will be frozen into dews.

(   ) 3. The northeast region enters winter as early as September. In the late period of October, Beijing has been influenced vividly by winter.

(   ) 4. Heavy Snow means the times of snowfalls increase, and the quantity of snow becomes large.

(   ) 5. Winter Solstice has the shortest daytime and longest nighttime in north hemisphere all the year around.

Ⅱ. Decide whether the statements are true (T) or false (F) according to the passage.

(   ) 1. Beginning of Autumn is a transition time from hotness to coolness. So the weather is not so stable, some days may be quite hot, some may be lower than expectations.

(   ) 2. White Dew is the real beginning of the cool autumn.

(   ) 3. Additionally, during Slight Cold, the rainfall is less than anytime else, and the weather is quite dry.

(   ) 4. According to the climatic data of China, the Great Cold is the phase with the lowest temperature.

(   ) 5. In Foshan, Guangdong Province, the local has the custom of braising the sticky rice. In Anqing, Anhui Province, the local has the custom of frying chunjuan.

Development

Ⅲ. Discuss the following questions.

1. Why do some areas not enter winter by the arrival of Beginning of Winter?

2. What does it mean that the Slight Cold is the phase with the lowest temperature?

第八章 农耕时令——七十二物候

Chapter 8 Farming Time—The 72 Phenological Terms

 七十二物候
The 72 Phenological Terms

　　Evolved on 24 solar terms, 72 phenological terms is the earliest calendar in China that combines astronomy, meteorology and phenology to guide agricultural activities. The ancients divided a year into 24 solar terms, each of which is divided into 3 phenological terms, a total of 72, reflecting the phenology of the year. Phenology is a phenomenon in which biotics and abiotics are affected by climate and other environmental factors, such as sprouting, spreading leaves, flowering, fruiting, insects and migratory birds coming and going, as well as frost, snow, thunder and water.

　　In *Xia Xiao Zheng*, the earliest agricultural calendar, 68 phenological phenomena, such as birds, animals, fish, insects, and abiotic , are mentioned as well as 7 kinds of climatic phenomena, 11 kinds of agricultural and animal husbandry on a monthly basis. As far as we know, the initial phenological records in China come from "sing of insects in May" in *The Book of Songs in July*, as well as "to peel dates in August", "to harvest rice in October". At that time, after carefully observation of phenology and accumulation of experience, the dates to sow, collection and harvest are chosen. These experiences are beneficial to the various production activities such as agriculture, forestry, animal husbandry by-fishing. During the Warring States period to the Western Han Dynasty, one year was divided into 72 phenological terms and 5 days as one phenological term in the book named *Yi Zhou Shu, Instruction of Seasons*, in which the phenological changes were observed and recorded for the first time. It was recorded 5 days as one phenological term, 3 phenological terms as one solar term, 6 solar terms as one season, 4 seasons as one year, a total of 24 solar terms, 72 phenological terms. Each phenological term is corresponding to one phenological phenomenon. The sequential changes in the 72 phenological terms reflect the climate changes of the year.

The Chinese calendar is a solar-lunar calendar—the solar calendar and lunar calendar. It integrates as well the revolution of the Earth around the Sun as the movement of the Moon around the Earth.

According to the solar calendar (the Gregorian Calendar), the dates of the solar terms is relatively fixed, with variation not exceeding three days, while the date on the lunar calendar varies according to the years.

The Chinese Lunar Calendar is a yearly one with the start of the lunar year based on the cycles of the moon. A normal year has 12 lunar months with about 30 days in each month, and during the leap year an extra month is added. That's to say, there are thirteen months in the leap year. A lunar month begins at the day of the new moon (invisible Moon) and ends at the day before the next new moon. The full moon is either on 15 or 16 of the month.

The Chinese Lunar Calendar has great influence on laborers life and agricultural production. Nowadays, the Chinese have officially adopted the Gregorian year or the solar calendar for the administration purpose, since the foundation of the People's Republic of China on October 1st, 1949.

Nevertheless, the Chinese People pay more attention to the lunar calendar. For example, almost all the Chinese People keep their traditional feasts fixed on the dates of the Chinese Lunar Calendar, such as the Spring Festivities, symbol of the arrival of the Chinese New Year. It always takes place on the first day of the first month on the Chinese Lunar Calendar.

Usually farmers keep the Chinese lunar calendar and look on the Beginning of Spring as the first solar term of a year. They often start a new round of ploughing after the Beginning of Spring.

72 phenological terms is the germination of ancient agro-meteorology and phenology in order to cope with climate change and guide agricultural production through the observation of climate and phenology in one year. The function of phenology mainly lies in forecasting weather and season, and plays a guiding role in preparing agricultural activities in ancient times. The ancients said: "When you don't miss the time of farming, there will be a rich harvest." All living things and non-living things in nature are a unity. In order to survive, animals and plants have to adapt to a variety of changes in all external environmental conditions. The changing conditions are immediately and intuitively reflected through the phenomenon of plants, which is phenology. Phenology is a myriad of "natural instruments" that comprehensively record changes in all environmental conditions at any time and any place. When climate conditions change to a certain value, plants make quantitative changes to qualitative ones—starting to germinate, spreading leaves, or flowering, and so on.

The phenology is closely related to farming. Through the early and late growth of plants, we can predict the changes in climate, dryness and humidity, and the early or late arrival of the season, then infer the farming season. According to *The Book of Fan Shengzhi, Ploughing,* "When apricots bloom, to plough the infertile soil; when apricots wither, to plough again." In the ancient times the root of the ploughing lay in the interesting season, which meant the season was particularly significant agricultural production. In addition, phenology was also employed by rulers to control the farmers. As recorded in *Huai Nan Zi, Column Nine*, "The king makes sure when the rainy season arrives, farmers will repair the field; when the shrimps calls for the swallow to return, they will repair the roads; when the star is located in the middle of the sky, it is necessary to consolidate the storage and cut down the fuel-wood for the winter". It reflects the use of phenology by the ruling class to "prepare and replenish the country and benefit the people".

Words & Expressions

astronomy/ə'strɒnəmi/ *n.* 天文学
meteorology/ˌmiːtiə'rɒlədʒi/ *n.* 气象学
phenology/fə'nɒlədʒi/ *n.* 物候学
germinate/'dʒɜːmɪneɪt/ *vt.* 使发芽；使生长；*vi.* 发芽；生长
replenish/rɪ'plenɪʃ/ *v.* 补充，重新装满；补足（原有的量）

Notes

1. *Xia Xiao Zheng*：《夏小正》是中国现存最早的一部记录农事的历书，收录于西汉戴德汇编《大戴礼记》第 47 篇。在《隋书·经籍志》首次出现《夏小正》单行本。本历书可窥见先秦中原农业发展水平，保存了古代中国的天文历法知识。《夏小正》撰者无考。一般认为成书时间为战国时期、两汉之间。

2. *Yi Zhou Shu, Instruction of Seasons*：战国至西汉期间《逸周书·时训解》，首次将一年分为七十二候，五日为一候，以此观测记录了物候的变化。以五日为候，三候为气，六气为时，四时为岁，一年二十四节气，共七十二候。各候均以一个物候现象相应，称"候应"。

3. *The Book of Fan Shengzhi, Ploughing*：《氾胜之书·耕田篇》记载："杏始华荣，辄耕轻土弱土。望杏花落，复耕。"

4. *Huai Nan Zi, Column Nine*：《淮南子·主术训》的记载："故先王之政，四海之云而至修封疆，虾蟆鸣，燕降而达路除道。……昴中则收敛蓄积，伐薪木"。

Reading Comprehension

Ⅰ. Each piece of the following information is given in one of the paragraphs in the passage. Identify the paragraph from which the information is derived and put the corresponding number in the space provided.

(   ) 1. The ancients divided a year into 24 solar terms, each of which is divided into 3 phenological terms, a total of 72, reflecting the phenology of the year.

(   ) 2. These experiences are beneficial for the various production activities such as agriculture, forestry, animal husbandry by-fishing.

(   ) 3. According to the solar calendar (the Gregorian Calendar), the dates of the solar terms is relatively fixed, with variation not exceeding three days, while the date on the lunar calendar varies according to the years.

(   ) 4. The Chinese Lunar Calendar has great influence on laborers life and agricultural production.

(   ) 5. 72 phenological terms is the germination of ancient agro-meteorology and phenology in order to cope with climate change and guide agricultural production through the observation of climate and phenology in one year.

Ⅱ. Decide whether the statements are true (T) or false (F) according to the passage.

(   ) 1. It was recorded 5 days as one phenological term, 3 phenological terms as one solar term, 6 solar terms as one season, 4 seasons as one year, a total of 24 solar terms, 72 phenological terms.

(   ) 2. A normal year has 12 lunar months with about 31 days in each month, and during the leap year an extra month is added.

(   ) 3. Usually farmers keep the Chinese lunar calendar and look on the Beginning of Spring as the first solar term of a year.

(   ) 4. The ancients said: "Even if you don't miss the time of farming, a rich harvest will be assured."

(   ) 5. The changing conditions are not immediately and intuitively reflected through the phenomenon of plants, which is phenology.

Language Focus

Ⅲ. Complete the sentences with the correct form of the words in the table.

accumulate	gregorian	intuitive	quantity	humid
abiotics	fertile	replenish	meteorology	germinate

1. Evolved on 24 solar terms, 72 phenological terms is the earliest calendar in China that combines astronomy, _____ and phenology to guide agricultural activities.

2. Phenology is a phenomenon in which biotics and _____ are affected by climate and other environmental factors:

3. At that time, after carefully observation of phenology and _____ of experience, the dates to sow, collection and harvest are chose.

4. According to the solar calendar (the _____ Calendar), the dates of the

solar terms is relatively fixed, with variation not exceeding three days, while the date on the lunar calendar varies according to the years.

5. The 72 phenological terms is the _____ of ancient agro-meteorology and phenology in order to cope with climate change and guide agricultural production through the observation of climate and phenology in one year.

6. The changing conditions are immediately and _____ reflected through the phenomenon of plants, which is phenology.

7. When climate conditions change to a certain value, plants make _____ changes to qualitative ones-starting to germinate, spreading leaves, or flowering, and so on.

8. Through the early and late growth of plants, we can predict the changes in climate, dryness and _____, and the early or late arrival of the season, then infer the farming season.

9. When apricots bloom, to plough the _____ soil; when apricots wither, to plough again.

10. It reflects the use of phenology by the ruling class to "prepare and_____ the country and benefit the people".

Ⅳ. Match the sentences in Section A with the English translation in Section B.

Section A

1. 据《氾胜之书·耕田篇》记载："杏始华荣，辄耕轻土弱土。望杏花落，复耕。"

2. 物候是生物与非生物受气候及其他环境因素影响而出现的现象，如草木发芽、展叶、开花、结实，昆虫和候鸟来去以及霜、雪、雷及结水等现象。

3. 根据植物生长发育的早迟，可知气候冷暖、干湿的变化，根据季节来临的早迟，推知农事季节。

4. 七十二物候是古代农业气象学和物候学的萌芽，通过观测一年中气候和物候现象以应对气候变化和指导农业生产。

5. 七十二物候是基于二十四节气发展而来，是我国最早的结合天文、气象、物候知识指导农事活动的历法。

Section B

1. The 72 phenological terms is the germination of ancient agro-meteorology and phenology in order to cope with climate change and guide agricultural production through the observation of climate and phenology in one year.

2. Evolved on 24 solar terms, 72 phenological terms is the earliest calendar in China that combines astronomy, meteorology and phenology to guide agricultural activities.

3. According to *The Book of Fan Shengzhi, Ploughing*, "When apricots bloom, to plough the infertile soil; when apricots wither, to plough again."

4. Phenology is a phenomenon in which biotics and abiotics are affected by climate and other environmental factors: such as sprouting, spreading leaves, flowering, fruiting, insects and migratory birds coming and going, as well as frost, snow, thunder and water.

5. Through the early and late growth of plants, we can predict the changes in climate, dryness and humidity, and the early or late arrival of the season, then infer the farming season.

V. Translate the paragraph into Chinese.

During the Warring States period to the Western Han Dynasty, one year was divided into 72 phenological terms and 5 days as one phenological term in the book named *Yi Zhou Shu, Instruction of Seasons*, in which the phenological changes were observed and recorded for the first time. It was recorded 5 days as one phenological term, 3 phenological terms as one solar term, 6 solar terms as one season, 4 seasons as one year, a total of 24 solar terms, 72 phenological terms. Each phenological term is corresponding to one phenological phenomenon.

_____
_____
_____
_____
_____
_____

VI. Discuss the following questions.

1. What is the function of phenology?
2. How is the phenology closely related to farming?

Text B 七十二物候解说
The Exact Interpretations of 72 Phenological Terms

Beginning of Spring is divided into three periods. In the first five-day period of Beginning of Spring, the cold weather is thawed by the warm east wind; in the second five-day period, ground beetles revive; and in the third five-day period, it seems that the swimming fishes carry floating ices on their backs. The change of feature of the solar term, Beginning of Spring, is seen clearly through the names of the three periods.

In the first five-day period of Beginning of Spring, the cold weather is thawed by the warm east wind. The cold winter is gradually blown away by east wind. And the ground is gradually thawed from the surface. A proverb says: "The spring wind cracks the tree barks", which means the spring wind is strong enough to crake the tree barks. A line in the poem, by the great Tang poet, Li Bai, "The spring wind breaks the glazed tile", vividly describes the strength of the spring wind. In spring, people do lots of work to prevent sandstorms. In south China, the verdant willow is bred by the strong and moist southeast wind. In such beautiful surroundings, poets created wonderful lines of poems like "The wind of early spring is sharp as scissor blades". Warmed by spring wind, spring returns to the earth, and the ground begins to melt.

In the second five-day period of Beginning of Spring, ground beetles revive. In this time, insects wake up from their long winter sleep, when they feel that the ground begins to be warm. Their bodies change to flexible from frozen. With the rising of temperature, insects, which hide underground, will gradually wake up from hazy sleep, and will raise up the spirit, devoting into new activities.

The third five-day period of Beginning of Spring, the swimming fishes are like carrying ices on the surface of water, because the ice has not been melted in this time. When Beginning of Spring is coming, fishes, hiding in deep water in winter, can realize the changes of water temperature after a few days. Although the surface of water starts to melt, there is a slice of thin ice on it, which prevents fishes to breathe fresh air. So fishes, swimming under the thin ice, are eager to feel this warm weather. Fishes, though blocked by ice, can enjoy the radiant and enchanting sunshine. And then, people aware that fishes swim under the water, and they define the situation as the signal to tell the arrival of spring.

Rain Water can be divided into three periods. During the first five-day period of Rain Water, bank beavers will catch fishes displayed like sacrificial offerings. Bank beaver is a small carnivorous animal with dense and silky fur, which is good at catching fishes in water. On Rain Water, ice begins to melt, and fishes swim on the surface of water. It is a good opportunity for bank beaver to catch them. After a good rest of the whole winter, fishes are stout and strong. When bank beavers catch fishes, they display fishes on the back, which is like they make sacrificial offerings.

The second five-day period of Rain Water means wild geese fly to north. Wild geese are migrants. They pass the winter in south, and return north in summer. In this time, weather begins to be warm in northern China, so flocks of geese will be seen flying in a line or in V shape formation to the north, seeking for their last home. When you see this situation, it means that Rain Water is coming.

In the third five-day period of Rain Water, grasses and plants will back to life. The temperature rises over 0 Degree Celsius, and spring rain begins to nurture the earth. Some early spring plants have already begun to grow. When you wander along the road in side or around the houses, you will notice that various plants are pushing out new shoots.

There are three periods of the Waking of Insect. The first five-day period of Waking of Insects means peach begins to blossom. In farmlands and yards, there are peach blossoms everywhere, in pink, white, and red. Some yards are filled with peach blossoms, and you can smell fragrance in farmland. Bright flowers make the ground looks lively. It is a splendid landscape to add some green grasses, trees and winter wheats.

In second five-day period of Waking of Insects, yellow birds will sing. Yellow birds are also called as Cang Geng. Trees and peach blossoms are thriving in the spring. It is comfortable for birds to go out in warm spring. Free yellow birds jump here and there in flowers and trees, and sing at their sweet will. Their beautiful and melodious songs win popularity with scholars and add pleasures to nature.

Eagles are replaced by turtledoves in the third five-day period of Waking of Insects. Two animals are mentioned, including eagles and turtledoves. Owing to the growth of plants on the ground, rabbits and field mice have places to hide. So you can hardly find eagles preying in sky, instead of turtledoves singing on trees.

The Spring Equinox has three periods. The first five-day period of the Spring Equinox means the black bird is coming. The black bird is the swallow, belonging to seasonal migratory bird. In the Spring Equinox, the weather is warm due to the south wind in north China. Swallows, pass winter in south China, fly back, and build house again. They keep grasses and mud in mouth, flying cross houses, trees and farmlands. Some swallows sing on the beams, and bring lots of fun to spring lives. And some swallows move to a new house. There is a poem said, "Swallows that nested under the eaves of dignitaries are flying to the homes of ordinary people. ("王" refers to "Wang Dao", and "谢" means "Xie An". They both are lord officials in the Jin Dynasty.)"

The second five-day period of the Spring Equinox means that the thunder is coming. When the weather is warm, the spring rains is frequent, and the air is wet. So the roaring of thunder rise from the distance. Du Xunhe, a famous poet in the Tang Dynasty, wrote a wonderful poem, *Eight Rhyme of Enjoying Mountain Pavilion with Friends*, said: "I read aloud with friend, and insects think it is a thunder as a mistake."

The third five-day period of the Spring Equinox indicated there is the first lightening. The air is wet in the Spring Equinox, and rains are growing concentrated. With thunder roaring everywhere, the lightening also happens. At this time, people can see lightening across the clouds, and hear the roaring thunder. As it described in mythology, the gods of wind, rain, thunder, and lightening gather together, spiriting the spring.

Pure Brightness can be divided into three periods. In the first five-day period of Pure Brightness, tung tree begins to have light purple flowers. Cluster of flowers, and lines of trees, bring beautiful spring to people.

In second five-day period of Pure Brightness, peony starts to blossom. Peony, the national beauty and heavenly fragrance, presenting the wealth, has flowered. They fight for gorgeous, tender and lovely, with bright colors. The most famous peony is in Luoyang, Henan Province, where holds Peony Festival every year. Louyang is a good place for people to outing.

In the third five-day period of Pure Brightness, rainbows can be observed after the rain. After Tomb-sweeping Festival, it rains frequently, with higher humidity in the air. If somewhere rains, occurring shining sun, the tiny drop of water will cast the rainbows. The beauty of the scenery defies all description.

Words & Expressions

thaw/θɔː/v. （使）（冰雪）融化；（使）（冷冻食品）解冻；使（身体等）变得暖起来；变得友好，变得缓和

crack/kræk/v. 破裂，裂开；崩溃，垮掉；砸开，砸碎；破译，解决；重击，猛击；（使）发出爆裂声，劈啪作响；（嗓音）变嘶哑；说（笑话），开（玩笑）；阻止，打击；开瓶，（尤指）开瓶饮酒；使（碳氢化合物）裂化

verdant/'vɜːd(ə)nt/adj. 青翠的；翠绿的；没有经验的；不老练的

hazy/'heɪzi/adj. 雾蒙蒙的，朦胧的；记不清的，模糊的；主意不定的，困惑的

radiant/'reɪdiənt/adj. 容光焕发的，喜悦的；明亮的，灿烂的；（热、能量）辐射的；（情感、品质）强烈的，显著的

enchanting/ɪn'tʃæntɪŋ/adj. 迷人的；妩媚的

carnivorous/kɑː'nɪvərəs/adj. 食肉的；肉食性的

stout/staʊt/adj. （人）肥胖的，壮实的；厚而结实的；不屈不挠的，顽强的

melodious/mə'ləʊdiəs/adj. 悦耳的，动听的；旋律优美的

turtledove/'tɜːtldʌv/n. 斑鸠

dignitary/'dɪɡnɪtəri/n. 高官；高僧；显要人物

tung/tʊŋ/n. 桐；桐树

humidity/hjuː'mɪdəti/n. 潮湿，湿气；湿度

Notes

1. Beginning of Spring is divided into three periods.

立春节气的 15 天分为三候，初候"东风解冻"，二候"蛰虫始振"，三候"鱼涉负冰"。说的是东风送暖，大地开始解冻；五日后，蛰居的虫类慢慢在洞中苏醒；再过五日，河里的冰开始融化，鱼开始到水面上游动，此时水面上还有没完全融解的碎冰片，如同被鱼负着一般浮在水面。

2. Rain Water can be divided into three periods.

雨水节分为三候，初候"獭祭鱼"，二候"候雁北"，三候"草木萌动"。这时候水獭开始捕鱼了，还要做出"先祭再食"的样子；五天后，大雁开始从南方飞回北方；再过五天，就能看到春雨霏霏、草木吐芽的早春气息了。

3. There are three periods of the Waking of Insect.

惊蛰节气分为三候，初候"桃始华"，二候"仓庚鸣"，三候"鹰化为鸠"。意思是说桃花自此渐盛；五日后黄鹂开始鸣叫；又五日，鹰开始悄悄地躲起来繁育后代，而原来蛰伏的鸠开始鸣叫求偶。

4. The Spring Equinox has three periods.

春分三候，初候"玄鸟至"，二候"雷乃发声"，三候"始电"。意思是说春分日后，燕子便从南方飞回来了，下雨时天空便打雷并发出闪电。

5. Pure Brightness can be divided into three periods.

清明节分为三候，初候"桐始华"，二候"牡丹华"，三候"虹始见"。在这个时节先是白桐花开放，接着喜阴的田鼠不见了，全回到了地下的洞中，然后是雨后的天空可以见到彩虹了。

Reading Comprehension

Ⅰ. **Each piece of the following information is given in one of the paragraphs in the passage. Identify the paragraph from which the information is derived and put the corresponding number in the space provided.**

(   ) 1. The cold winter is gradually blown away by east breeze.

(   ) 2. The second five-day period of Rain Water means wild geese fly to north.

(   ) 3. Trees and peach blossoms are thriving in the spring.

(   ) 4. When the weather is warm, the spring rains is frequent, and the air is wet. So the roaring of thunder rise from the distance.

(   ) 5. After Tomb-sweeping Festival, it rains frequently, with higher humidity in the air.

Ⅱ. **Decide whether the statements are true (T) or false (F) according to the passage.**

(   ) 1. Fishes, blocked by ice, cannot enjoy the radiant and enchanting sunshine.

(   ) 2. Wild geese pass the winter in north, and return south in summer.

(   ) 3. Owing to the growth of plants on the ground, rabbits and field mice have places to hide. So you can hardly find eagles preying in sky, instead of turtledoves singing on trees.

(   ) 4. In the Spring Equinox, the weather is warm due to the south wind in north China.

(   ) 5. In first five-day period of Pure Brightness, peony starts to blossom.

Development

Ⅲ. **Discuss the following questions.**

1. What are the three phenological terms for Beginning of Spring?
2. In which phenological term peach begins to blossom?

物候学
Phenology

Phenology is the study of the timing of the biological events in plants and animals such as flowering, leafing, hibernation, reproduction, and migration. Scientists who study phenology are interested in the timing of such biological events in relation to changes in season and climate.

How does a bear know when it's time to hibernate? Why do April showers bring May flowers? Plants and animals don't have calendars or watches, but many of them take cues from the changing seasons. Changes in weather with the seasons, such as temperature and precipitation, signal many organisms to enter new phases of their lives. For example, buds form on plants as temperatures warm in the spring. As temperatures cool in the fall, deciduous trees and shrubs lose their leaves and become dormant. The study of the timing of these changes is called phenology.

Phenology is literally "the science of appearance". The word phenology comes from the Greek words phaino (to show or appear) and logos (to study). Scientists who study phenology—phenologists—are interested in the timing of specific biological events (such as flowering, migration, and reproduction) in relation to changes in season and climate. Seasonal and climatic changes are some of the non-living or abiotic components of the environment that impact the living or biotic components. Seasonal changes can include variations in day length, temperature, and rain or snowfall. In short, phenologists attempt to learn more about the abiotic factors to which plants and animals respond.

Examples of springtime phenological events that interest scientists include flowering, leaf unfolding, insect emergence, and bird, fish, and mammal migration. The arrival of spring gets a lot of attention in terms of phenological events, with flowers emerging from their winter slumber. However, equally important phenological events happen throughout the year.

The timing of when a place warms up in the spring and when it cools down in the fall depends on Earth's climate. Regional differences in climate cause warm weather to arrive later in the spring at higher latitudes than at lower latitudes. But today global change in climate is affecting the timing of warming temperatures in the spring and cooling temperatures in the fall worldwide.

As Earth's temperature rises, it becomes warmer earlier in the spring and stays warmer later into the fall at any given location. The opposite is true as well: if Earth's temperature were to cool, warm weather would show up later in the spring and cool weather would arrive earlier in the fall. Today, the global climate is warming.

Changes in the timing of phases of the plant life cycle, known as phenophases, are directly affected by temperature, rainfall, and day length. While these factors change through the year in places where there are distinct seasons, the first two — temperature and rainfall — are also changing in many regions because of climate change. For example, if climate change causes warmer temperatures, warm weather may occur earlier in the spring and it may stay warm later into the fall than in years past. It will still get cold in the winter and warm in the summer, but the plant growing season will be longer, and that can have big impacts on living things.

That's where plants come in. By monitoring plants and noting when the first buds appear, when the first flowers appear, when leaves drop in the fall, and other parts of plant life cycles, scientists can figure out how seasonal patterns are changing, and make predictions for the future.

Phenological observations have been used for centuries by farmers to maximize crop production, by nature-lovers to anticipate optimal wildflower viewing conditions, and by many of us to prepare for seasonal allergies.

Changes in phenological events can have a significant impact on how we humans live our lives and interact with our environment on a daily basis.

Having a parade for cherry blossoms while the blooms are fading is bad timing, but it is perhaps not quite as dire as some cases of bad timing that affects entire ecosystems. For example, in most ecosystems, there are insects and plants that need each other. Hungry insects searching for nectar from flowers inadvertently transport pollen from flower to flower. The pollen grains hitch a ride, often by sticking to an insect's legs. By distributing pollen, the insects, called pollinators, are fertilizing the flowers, allowing the plant to grow seeds and fruit.

But it takes time for insects to develop from egg to larva to adult, and the timing of their growth can't be sped up just because the flowers are blooming earlier. As the climate warms, plants may become out of sync with the insects that pollinate them. If an insect is still a larva when the flowers blossom, for example, it will not be able to fly from flower to flower to transport pollen. Without pollination, the flowers will not be fertilized and will not produce fruit.

Mammals in the ecosystem can be affected, too. Consider mice: some mice eat insects and seeds. If plants bloom too early for insects to pollinate them, then the seeds won't grow. And if the insects are too late to gather food from the flowers, they will not survive, either. Without seeds or insects to eat, the mice may not survive. And animals that eat mice, like snakes and hawks, will also go hungry.

How plants react to seasonal change has a big impact on the natural environment. Because plants are at the base of the food chain, anything that affects plants can impact other parts of the ecosystem. Phenology is important because it affects whether plants and animals thrive or survive in their environments. It is important because our food supply depends on the timing of phenological events. And, to scientists, changes in the timing of phenological events can be used an as indicator of changing climates.

Changes in phenological events can also have a significant impact on how we humans live our lives and interact with our environment on a daily basis. For example, the timing of when plants flower and fruit can affect our food supply and therefore our health. Pollen allergies can also be exacerbated by changes in growing conditions. People who are allergic to plant pollen will experience reactions to the changes in flowering times and the lengthening of the growing season.

From historical records and observations, we know that phenological events can vary from year to year. Ecosystems can recover from variation between years, but when these changes happen consistently over many years, the timing of events such as flowering, leafing, insect emergence, and allergies can impact how plants, animals, and humans are able to thrive in their environments.

Words & Expressions

hibernation /ˌhaɪbə'neɪʃn/ n. 冬眠；（人）蛰居，不活动
precipitation /prɪˌsɪpɪ'teɪʃn/ n. 降水
deciduous /dɪ'sɪdʒuəs/ adj. 落叶性的，脱落性的
shrub /ʃrʌb/ n. 灌木
abiotic /ˌeɪbaɪ'ɒtɪk/ adj. 非生物的，无生命的
slumber /'slʌmbə(r)/ n. 睡眠；麻木状态；静止状态
phenophase /'fiːnəfeɪs/ n. 物候期；物候相
nectar /'nektə(r)/ n. 花蜜
exacerbate /ɪɡ'zæsəbeɪt/ vt. 使加剧；使恶化

Notes

Phenology：物候学，是对周期性生物现象的科学研究，如与气候状况相关的植物开花、动物迁移。

Reading Comprehension

I. Each piece of the following information is given in one of the paragraphs in the passage. Identify the paragraph from which the information is derived and put the corresponding number in the space provided.

(　　) 1. Phenology is literally "the science of appearance." The word phenology comes from the Greek words phaino (to show or appear) and logos (to study).

(　　) 2. The timing of when a place warms up in the spring and when it cools down in the fall depends on Earth's climate.

(　　) 3. Changes in the timing of phases of the plant life cycle, known as phenophases, are directly affected by temperature, rainfall, and day length.

(　　) 4. Mammals in the ecosystem can be affected, too.

(　　) 5. Changes in phenological events can also have a significant impact on how we humans live our lives and interact with our environment on a daily basis.

II. Decide whether the statements are true (T) or false (F) according to the passage.

(　　) 1. A bear know when it's time to hibernate even if he doesn't have calendars.

(　　) 2. Seasonal and climatic changes are some of the non-living or abiotic components of the environment that impact the living or biotic components.

(　　) 3. Seasonal differences in climate cause warm weather to arrive later in the spring at higher latitudes than at lower latitudes.

(　　) 4. If climate change causes warmer temperatures, warm weather may occur earlier in the spring and it may stay warm later into the fall than in years past. So the plant growing season will be shorter.

(   ) 5. Without pollination, the flowers will not be fertilized and will not produce fruit.

Development

III. Discuss the following questions.

1. Why is phenology important?

2. How do insects and plants need each other?

参考文献
Reference

宋先道，2011．中国市场与文化［M］．武汉：湖北科学技术出版社．
田华实，仲锡，2008．用英语说中国——艺术［M］．上海：上海科学普及出版社．
王德军，吕芸芳，2008．用英语说中国——文化［M］．上海：上海科学普及出版社．
吴圣正，2019．中国传统文化概说［M］．北京：人民出版社．
苑荣，唐志强，2018．五千年农耕的智慧：中国古代农业科技知识［M］．北京：中国农业出版社．
曾小荣，2015．中国农业史英语读本［M］．北京：外语教学与研究出版社．
About Phenology [EB/OL]. (2019-10-25)/[2019-11-30]. http://budburst.org/phenology.